Biotechnology for Agricultural Breeding

Biotechnology for Agricultural Breeding

Biotechnology for Agricultural Breeding

Editor
S.K. Mangal

2014

Gene-Tech Books
New Delhi - 110 002

ISBN 81-89729-27-6

Published by : **GENE-TECH BOOKS**
4762-63/23, Ansari Road, Darya Ganj,
NEW DELHI - 110 002
Phone: 41562849
e-mail: genetechbooks@yahoo.co.in

Printed at : **Tarun Offset Printers**
Delhi

PRINTED IN INDIA

Preface

Agricultural biotechnology is an advanced technology that allows plant breeders to make precise genetic changes to impart beneficial traits to the crop plants we rely on for food and fibre. For centuries farmers and plant breeders have laboured to improve crop plants. Traditional breeding methods include selecting and sowing the seeds from the strongest, most desirable plants to produce the next generation of crops. By selecting and breeding plants with characteristics such as higher yield, resistance to pests and hardiness, early farmers dramatically changed the genetic make-up of crop plants long before the science of genetics was understood. As a result, most of today's crop plants bear little resemblance to their wild ancestors.

The tools of modern biotechnology allow plant breeders to select genes that produce beneficial traits and move them from one organism to another. This process is far more precise and selective than crossbreeding, which involves the transfer of tens of thousands of genes, and provided plant developers with a more detailed knowledge of the changes being made. Depending on which genes are transferred, agricultural biotechnology can protect crops from disease, increase their yield, improve their nutritional content, or reduce pesticide use. By enhancing the nutritional value of foods, biotechnology can help improve the quality of basic diets. Agricultural biotechnology also provides benefits for the manufacture of pharmaceutical products. Because plants do not carry human diseases, plant-made vaccines and antibodies require less screening for bacterial toxins and viruses. In addition to plants, animals may also be engineered to produce beneficial genes. In order to produce large quantities of monoclonal antibodies for research on new therapeutic drugs, several companies have genetically engineered cows and goats to secrete antibodies into their milk.

Despite the benefits of genetic engineering, there are concerns about whether recombinant DNA techniques carry greater risks than traditional breeding methods. Consumer acceptance of food derived from genetically engineered crops has been variable. Many scientists express concerns regarding the environmental impact and ethics of the new technology, and about food safety.

This book provides a comprehensive introduction to the applications of biotechnology in agriculture. It integrates basic biotechnological methodologies with up-to-date agricultural practices, offering solutions to specific agricultural needs and problems from plant and crop yield to animal husbandry. It evaluates the limitations of classical methodologies and the potential of novel and emergent agriculturally related biotechnologies. It will surely be a valuable reference tool for students, libraries and active research workers.

Editor

Contents

1

Agricultural Biotechnology

Biotechnology can be applied to all classes of organism—from viruses and bacteria to plants and animals—and it is becoming a major feature of modern medicine, agriculture and industry. Modern agricultural biotechnology includes a range of tools that scientists employ to understand and manipulate the genetic make-up of organisms for use in the production or processing of agricultural products. Some applications of biotechnology, such as fermentation and brewing, have been used for millennia. Other applications are newer but also well established.

For example, micro-organisms have been used for decades as living factories for the production of life-saving antibiotics including penicillin, from the fungus Penicillium, and streptomycin from the bacterium Streptomyces. Modern detergents rely on enzymes produced via biotechnology, hard cheese production largely relies on rennet produced by biotech yeast and human insulin for diabetics is now produced using biotechnology.

Biotechnology is being used to address problems in all areas of agricultural production and processing. This includes plant breeding to raise and stabilise yields; to improve resistance to pests, diseases and abiotic stresses such as drought and cold; and to enhance the nutritional content of foods. Biotechnology is being used to develop low-cost disease-free planting materials for crops such as cassava, banana and potato and is creating new tools for the diagnosis and treatment of plant and animal diseases and for the measurement and conservation of genetic resources.

Biotechnology is being used to speed up breeding programmes for plants, livestock and fish and to extend the range of traits that can be addressed. Animal feeds and feeding practices are being changed by biotechnology to improve animal nutrition and to reduce environmental waste. Biotechnology is used in disease diagnostics and for the production of vaccines against animal diseases. Clearly, biotechnology is more than genetic engineering. Indeed, some of the least controversial aspects of agricultural biotechnology are potentially the most powerful and the most beneficial for the poor.

Genomics, for example, is revolutionising our understanding of the ways genes, cells, organisms and ecosystems function and is opening new horizons for marker-assisted breeding and genetic resource management. At the same time, genetic engineering is a very powerful tool whose role should be carefully evaluated. It is important to understand how biotechnology—particularly genetic engineering—complements and extends other approaches if sensible decisions are to be made about its use.

The Convention on Biological Diversity (CBD) defines biotechnology as: "any technological application that uses biological systems, living organisms, or derivatives thereof, to make or modify products for specific use". This definition includes medical and industrial applications as well as many of the tools and techniques that are commonplace in agriculture and food production. The Cartagena Protocol on Biosafety defines "modern biotechnology" more narrowly as the application of:

a) In vitro nucleic acid techniques, including recombinant deoxyribonucleic acid (DNA) and direct injection of nucleic acid into cells or organelles, or

b) Fusion of cells beyond the taxonomic family, that overcome natural physiological reproductive or recombination barriers and that are not techniques used in traditional breeding and selection.

Recombinant DNA techniques, also known as genetic engineering or (more familiarly but less accurately) genetic modification, refer to the modification of an organism's genetic make-up using transgenesis, in which DNA from one organism or cell (the transgene) is transferred to another without sexual reproduction. Genetically modified organisms (GMOs) are modified by the application of transgenesis or recombinant DNA technology, in which a transgene is incorporated into the host genome or a gene in the host is modified to change its level of expression. The terms "GMO", "transgenic organism" and "genetically engineered organism (GEO)" are often used interchangeably although they are not technically identical.

Characterisation of Genetic Resources

Farmers and pastoralists have manipulated the genetic make-up of plants and animals since agriculture began more than 10,000 years ago. Farmers managed the process of domestication over millennia, through many cycles of selection of the best adapted individuals. This exploitation of the natural variation in biological organisms has given us the crops, plantation trees, farm animals and farmed fish of today, which often differ radically from their early ancestors (Table 1). The aim of modern breeders is the same as that of early farmers—to produce superior crops or animals.

Conventional breeding, relying on the application of classic genetic principles based on the phenotype or physical characteristics of the organism concerned, has been very successful in introducing desirable traits into crop cultivars or livestock breeds from domesticated or wild relatives or mutants. In a conventional cross, whereby each parent donates half the genetic make-up of the progeny, undesirable traits may be passed on along with the desirable ones, and these undesirable traits may then have to be eliminated through successive generations of breeding. With each generation, the progeny must be tested for its growth characteristics as well as its nutritional and processing traits.

Many generations may be required before the desired combination of traits is found, and time lags may be very long, especially for perennial crops such as trees and some species of livestock. Such phenotype-based selection is thus a slow, demanding process and is expensive in terms of both time and money. Biotechnology can make the application of conventional breeding methods more efficient.

Table 1. An agricultural technology timeline

Technology	Era	Genetic interventions
Traditional	About 10 000 years BC	Civilisations harvested from natural biological diversity, domesticated crops and animals, began to select plant materials for propagation and animals for breeding

contd...

contd...

	About 3 000 years BC	Beer brewing, cheese making and wine fermentation
Conventional	Late nineteenth century	Identification of principles of inheritance by Gregor Mendel in 1865, laying the foundation for classical breeding methods
	1930s	Development of commercial hybrid crops
	1940s to 1960s	Use of mutagenesis, tissue culture, plant regeneration. Discovery of transformation and transduction. Discovery by Watson and Crick of the structure of DNA in 1953. Identification of genes that detach and move (transposons)
Modern	1970s	Advent of gene transfer through recombinant DNA techniques. Use of embryo rescue and protoplast fusion in plant breeding and artificial insemination in animal reproduction
	1980s	Insulin as first commercial product from gene transfer. Tissue culture for mass propagation in plants and embryo transfer in animal production
	1990s	Extensive genetic fingerprinting of a

contd...

contd...

	wide range of organisms. First field trials of genetically engineered plant varieties in 1990 followed by the first commercial release in 1992. Genetically engineered vaccines and hormones and cloning of animals
2000s	Bioinformatics, genomics, proteomics, metabolomics

Structure of Genomics

The most significant breakthroughs in agricultural biotechnology are coming from research into the structure of genomes and the genetic mechanisms behind economically important traits. The rapidly progressing discipline of genomics is providing information on the identity, location, impact and function of genes affecting such traits—knowledge that will increasingly drive the application of biotechnology in all agricultural sectors. Genomics sets the foundation for post-genomics activities, including new disciplines such as proteomics and metabolomics to generate knowledge on gene and protein structure, as well as their functions and interactions. These disciplines seek to understand systematically the molecular biology of organisms for their practical use.

A vast range of new and rapidly advancing technologies and equipment has also been developed to generate and process information about the structure and function of biological systems. The use and organisation of this information is called bioinformatics. Advances in bioinformatics may allow the prediction of gene function from gene sequence data: from a listing of an organism's genes, it will become possible to build a theoretical framework of its biology. The comparison across organisms of physical and genetic maps and DNA sequences will significantly reduce the time needed to identify and select potentially useful genes.

Through the production of genetic maps that provide the precise location and sequences of genes, it is apparent that even distantly related genomes share common features. Comparative genomics assists in the understanding of many genomes based on the intensive study of just a few. For instance, the rice genome sequence is useful for studying the genomes of other cereals with which it shares features according to its degree of relatedness, and the mouse and malaria genomes provide models for livestock and some of the diseases that affect them. There are now model species for most types of crops, livestock and diseases and knowledge of their genomes is accumulating rapidly.

All living things are made up of cells that are programmed by genetic material called deoxyribonucleic acid (DNA). Only a small fraction of the DNA chain actually makes up genes, which in turn code for proteins, and the remaining share of the DNA represents non-coding sequences whose role is not yet clearly understood. The genetic material is organised into pairs of chromosomes. For example, there are five chromosome pairs in the much-studied mustard species Arabidopsis thaliana. An organism's entire set of chromosomes is called the genome.

The Human Genome Sequencing Project has provided the agricultural research community not only with many spin-off technologies that can be applied across the board for all living organisms but also with a model for international collaboration in tackling large genome-sequencing projects for model plants such as Arabidopsis and rice.

Molecular Markers

Reliable information on the distribution of genetic variation is a prerequisite for sound selection, breeding and conservation programmes. Genetic variation of a species or population can be assessed in the field or by studying molecular and other markers in the laboratory. A combination of the two approaches is required for reliable results. Molecular markers are identifiable DNA sequences, found at specific locations of the genome and associated with the inheritance of a trait or linked gene. Molecular markers can be used for (a) marker-assisted breeding, (b) understanding and conserving genetic resources and (c) genotype verification. These activities are critical for the genetic improvement of crops, forest trees, livestock and fish.

Marker-assisted Breeding

Genetic linkage maps can be used to locate and select for genes affecting traits of economic importance in plants or animals. The potential benefits of marker-assisted selection (MAS) are greatest for traits that are controlled by many genes, such as fruit yield, wood quality, disease resistance, milk and meat production, or body fat, and that are difficult, time-consuming or expensive to measure. Markers can also be used to increase the speed or efficiency of introducing new genes from one population to another, for example when wishing to introduce genes from wild relatives into modern plant varieties. When the desired trait is found within the same species, it may be transferred with traditional breeding methods, with molecular markers being used to track the desired gene.

Pearl millet is a cereal grown for foodgrain and straw in the hottest, driest areas of Africa and Asia where rainfed and dryland agriculture are practised. It is similar to maize in its breeding behaviour. Traditional farmers' varieties are open-pollinated and out-breeding and thus continuously changing. Genetically uniform hybrid varieties have been developed that offer higher yield potential but are more vulnerable to a plant disease called downy mildew. In India, pearl millet is grown on about 9 million ha and more than 70 percent of this is sown to such hybrid cultivars.

Since pearl millet hybrids first reached farmers' fields in India in the late 1960s, every variety that has become popular with farmers has ultimately succumbed to a downy mildew epidemic. Unfortunately, by the time the poorer farmers in a given region decide to adopt a particular variety, its days are usually numbered. The International Crops Research Institute for the Semi-Arid Tropics (ICRISAT) wanted to reduce the risks associated with adoption of higher-yielding pearl millet hybrids and extend the useful economic life of these varieties, especially for poorer producers.

Measuring Genetic Diversity

The use of molecular markers to measure the extent of variation at the genetic level, within and among populations, is of value in guiding genetic conservation activities and in the development of breeding populations in crops, livestock, forestry and fisheries. Studies carried out using these technologies in fish and forest tree species have revealed high levels of genetic variation both among and within populations. Livestock species are characterised by a high degree of genetic variation within populations, whereas crops exhibit a higher degree of variation across species.

Molecular markers are increasingly used to study the distribution and patterns of genetic diversity. Global surveys indicate, for example, that 40 percent of the remaining domestic livestock breeds are at risk of extinction. Most of these breeds are found only in developing countries, and there is often little knowledge about them or of their potential for improvement. They may contain valuable genes that confer adaptation or resilience to stresses, such as heat tolerance or disease resistance, that may be of use for future generations. Modern biotechnologies can help to counteract trends of genetic erosion in all food and agriculture sectors.

Genotype Verification

Molecular markers have been widely used for identifying genotypes and for "genetic fingerprinting" of organisms. Genetic fingerprinting has been used in advanced tree-breeding programmes in which the correct identification of clones for large-scale propagation programmes is essential. Molecular markers have been used to identify endangered marine species that are either inadvertently captured in wild fisheries or that are purposefully taken illegally. Genotype verification is used intensively in parentage testing of domestic animals and for tracing livestock products in the food chain back to the farm and animal of origin.

Breeding and Reproducing Crops and Trees

A number of biotechnologies are used in breeding and reproducing crops and trees. Often these technologies are used in combination with each other and with conventional breeding approaches.

Tissue Culture and Micropropagation

Micropropagation involves taking small sections of plant tissue, or entire structures such as buds, and culturing them under artificial conditions to regenerate complete plants. Micropropagation is particularly useful for maintaining valuable plants, breeding otherwise difficult-to-breed species (e.g. many trees), speeding up plant breeding and providing abundant plant material for research. For crop and horticultural species, micropropagation is now the basis of a large commercial industry involving hundreds of laboratories around the world. In addition to its rapid propagation advantages, micropropagation can also be used to generate disease-free planting material, especially if combined with the use of disease-detection diagnostic kits.

There have been some attempts to use micropropagation more widely in forestry. Compared with vegetative propagation through cuttings, the higher multiplication rates available through micropropagation offer a more rapid dissemination of planting stock, although limited availability of desirable clones is an impediment to its wider adoption in forestry.

For example, banana is generally grown in developing countries where it is a source of employment, income and food. Banana production is in decline in many regions because of pest and disease problems that cannot be addressed successfully through agrochemical control for reasons of cost and negative environmental effects. The problem is exacerbated because banana is reproduced clonally; the use of diseased mother plants therefore gives rise to diseased offspring.Micropropagation represents a means of regenerating disease-free banana plantlets from healthy tissue.

In Kenya, banana shoot tips have been successfully tissue-cultured. An original shoot tip is heat-treated to destroy infective organisms and then used through many cycles of regeneration to produce daughter plants. A single section of tissue can be used to produce as many as 1,500 new plants through ten cycles of regeneration.

Micropropagation of banana has had a tremendous impact in Kenya, among many other countries, contributing to improved food security and income generation. It has all the advantages of being a relatively cheap and easily applied technology and one that brings significant environmental benefits.

In vitro selection

In vitro selection refers to the selection of germplasm by applying specific selection pressure to tissue culture under laboratory conditions. Many recent publications have reported useful correlations between in vitro responses and the expression of desirable field traits for crop plants, most commonly disease resistance. Positive results are available also for tolerance to herbicides, metals, salt and low temperatures. For the selection criteria of major general importance in forest trees (in particular vigour, stem form and wood quality), poor correlations with field responses still limit the usefulness of in vitro selection. However, this method may be of interest in forestry programmes for screening disease resistance and tolerance to salt, frost and drought.

Genetic Engineering

When the desired trait is found in an organism that is not sexually compatible with the host, it may be transferred using genetic engineering. In plants, the most common method for genetic engineering uses the soil bacterium Agrobacterium tumefasciens as a vector. Researchers insert the desired gene or genes into the bacterium and then infect the host plant. The desired genes are transmitted to the host along with the infection. This method is used mainly with dicot species such as tomato and potato.

Some crops, particularly monocot species such as wheat and rye, are not naturally susceptible to transformation via A. tumefasciens, although the method has recently been successfully used to transform wheat and other cereals. In the most common transformation technique for these crops, the desired gene is coated on gold or tungsten particles and a "gene gun" is used literally to shoot the gene into the host at high velocity.

Three distinctive types of genetically modified crops exist: (a) "distant transfer", in which genes are transferred between organisms of different kingdoms (e.g. bacteria into plants); (b) "close transfer", in which genes are transferred from one species to another of the same kingdom (e.g. from one plant to another); and (c) "tweaking", in which genes already present in the organism's genome are manipulated to change the level or pattern of expression.

Once the gene has been transferred, the crop must be tested to ensure that the gene is expressed properly and is stable over several generations of breeding. This screening can usually be performed more efficiently than for conventional crosses because the nature of the gene is known, molecular methods are available to determine its localisation in the genome and fewer genetic changes are involved.

Most of the transgenic crops planted so far have incorporated only a very limited number of genes aimed at conferring insect resistance and/or herbicide tolerance. However, some transgenic crops and traits of greater potential interest for developing countries have been developed but have not yet been released commercially.

Nutritionally enhanced crops could make a significant contribution to the reduction of micronutrient malnutrition in developing countries. Biofortification (the development of nutritionally enhanced foods) can be advanced through the application of several biotechnologies in combination. Genomic analysis and genetic linkage mapping are needed to identify the genes responsible for natural variation in nutrient levels of common foods (Table 2). These genes can then be transferred into familiar cultivars through conventional breeding and MAS or, if sufficient natural variation does not occur within a single species, through genetic engineering. Non-transgenic approaches are being used, for example, to enhance the protein content in maize, iron in rice, and carotene in sweet potato and cassava.

Table 2.
Genetic variation in concentrations of iron, zinc, beta-carotene and ascorbic acid found in germplasm of five staple foods, dry weight basis

	(mg/kg)			
	Iron	Zinc	Beta-carotene[1]	Ascorbic acid
RICE				
Brown	6-25	14-59	0-1	-
Milled	1-14	14-38	0	-
CASSAVA				
Root	4-76	3-38	1-242	0-38020
Leaves	39-236	15-109	180-9602	17-42002
BEAN	134-111[1]	21-540	0	-
MAIZE	10-63	12-580	0-10	-
WHEAT	10-99[3]	000[2]8-177[2]	0-20	-

[1] Range for total carotenoids is much greater. [2] Fresh weight basis.[3] Including wild relatives.

Genetic engineering can be used when insufficient natural variation in the desired nutrient exists within a species. The well-known transgenic Golden Rice contains three foreign genes—two from the daffodil and one from a bacterium—that produce provitamin A. Scientists are well on their way to developing transgenic "nutritionally optimised"' rice that would contain genes producing provitamin A, iron and more protein. Other nutritionally enhanced foods are under development, such as oils with reduced levels of undesirable fatty acids. In addition, foods that are commonly allergenic (shrimp, peanuts, soybean, rice, etc.) are being modified to contain lower levels of allergenic compounds.

A major technical factor limiting the application of genetic modification to forest trees is the current low level of knowledge regarding the molecular control of traits that are of most interest. One of the first reported trials with genetically modified forest trees was initiated in Belgium in 1988 using poplars. Since then, there have been more than 100 reported trials involving at least 24 tree species, primarily timber-producing species. Traits for which genetic modification has been contemplated for forest trees include insect and virus resistance, herbicide tolerance and lignin content. Reduction of lignin is a valuable objective for species producing pulp for the paper industry because it would enable a reduction in the use of chemicals in the process.

Researchers at Jawaharlal Nehru University in India have developed a genetically engineered potato that produces about one-third to one-half more protein than usual, including substantial amounts of all the essential amino acids such as lysine and methionine. Protein deficiency is widespread in India and potato is the staple food of the poorest people.

The "protato" was developed by a coalition of Indian charities, scientists, government institutes and industry as part of a 15-year campaign against childhood mortality. The campaign aims to eliminate childhood mortality by providing children with clean water, better food and vaccines.

The protato includes a gene from the amaranth plant, a high-protein grain that is native to South America and widely sold in Western health-food stores. The protato has passed preliminary field trials and tests for allergens and toxins. Final approval from the Indian Government is probably at least five years away. Supporters such as Govindarajan Padmanaban, a biochemist at the Indian Institute of Science, argue that the protato can provide an important nutritional boost to children with little danger of allergy because potatoes and amaranth are both already widely consumed. There is also little threat to the environment because neither potatoes nor amaranth have wild relatives in India, and the protato does not involve any change in normal potato production practices.

Furthermore, because the protato was developed by public-sector scientists in India, there are no concerns about foreign corporate control of the technology. Given these benefits, Padmanaban commented: "I think it would be morally indefensible to oppose it". Opponents such as Charlie Kronick of Greenpeace argue that potatoes are naturally quite low in protein (about 2 percent), so even a doubling of the protein content would make only a minute contribution to India's malnutrition problem. He claims that the effort to develop the protato was aimed more at gaining public acceptance of genetic engineering than at addressing the problem of malnutrition: "The cause of hunger isn't lack of food. It's lack of cash and of access to the food. Creating these GM crops is something to make them look attractive when actually the utility of eating them is very, very low. It's very difficult to see how this on its own will change the face of poverty".

Livestock and Fish Breeding

Biotechnology has long been a source of innovation in

livestock and aquaculture production and processing and has had a profound impact on both sectors. Rapid advances in molecular biology and further developments in reproductive biology provide new and powerful tools for further innovation. Technologies such as genomics and molecular markers, as described above, are valuable in understanding, characterising and managing genetic resources in livestock and fisheries as well as in crops and forestry.

Genetic engineering is also relevant in livestock and fisheries, although the techniques differ, and additional reproductive technologies are available in these sectors. The reproductive biotechnologies that are specific to the livestock and fisheries sectors. The main objective of reproductive biotechnologies for livestock is to increase reproductive efficiency and rates of animal genetic improvement. The genetic improvement of locally adapted breeds will be important in realising sustainable production systems within the broad spectrum of developing country production environments, and will probably best be realised by the strategic use of both non-genetic and genetic interventions. Reproductive biotechnology in fisheries presents opportunities to increase growth rates and improve the management of farmed species and to limit the reproductive potential of genetically engineered species.

Artificial Insemination and Multiple Ovulation

Advances in artificial insemination (AI) and multiple ovulation followed by embryo transfer (MOET) have already had a major impact on livestock improvement programmes in developed countries and many developing countries because they speed up the process of genetic improvement, reduce the risk of disease transmission and expand the number of animals that can be bred from a superior parent—the male in the case of AI and the female in the case of MOET. They also increase the incentives for private research in animal breeding and significantly expand the market for improved parent stock.

The number of AIs performed globally during 1998 was over 100 million in cattle (primarily dairy cattle, including buffalo), 40 million in pigs, 3.3 million in sheep and 0.5 million in goats. These figures illustrate both the higher economic returns in dairy cattle and the fact that cattle semen is much easier to deep-freeze than semen from other animals. Although over 60 million cattle AIs were performed in South and Southeast Asia, fewer than 1 million were performed in Africa. AI is only effective when the farm sector has access to considerably greater technical and institutional and other organisational capacity than when male animals are used directly for breeding purposes. On the positive side, farmers employing AI do not have to face the costs or hazards of rearing breeding males and can have access to semen from any part of the world.

Despite the widespread use of AI in developed countries and in many developing countries, including within more advanced smallholder systems, it is applied only on farms that practise intensive livestock management with high-value animals. This is clearly not because of technical problems with semen production and storage, as most procedures are now fully standardised and proven to be effective under tropical developing-country conditions. Rather, it is because of the many organisational, logistical and farmer-training constraints that influence the quality and efficiency of the technology.

MOET takes AI one step further, both in terms of genetic gains possible and level of technical capacity and organisation required. MOET is one of the basic technologies for the application of more advanced reproductive biotechnologies such as cloning and transgenics. During 2001 the number of embryos transferred globally was 450 000, mainly in dairy cattle, with North America and Europe accounting for 62 percent, followed by South America (16 percent) and Asia (11 percent). About 80 percent of the bulls used in AI are derived from MOET. The main potential advantage of MOET for developing countries will be in the possibility of importing frozen embryos instead of live animals, for example in the establishment of nucleus breeding stocks of locally adapted genetic resources, with the related lower sanitary risks.

Set Manipulation and Sex Reversal in Fish

Controlling the sex and reproductive capacity of fish can be important for commercial and environmental reasons. One sex is often more desirable than the other; for example, only female sturgeon produce caviar and male tilapia grow faster than females. Sterility may be desirable when reproduction affects the taste of the product (e.g. oysters) or when farmed species (transgenic or not) might breed with wild populations. Chromosome-set manipulation and sex reversal are well-established techniques to control these factors.

In chromosome-set manipulation, temperature, chemical and pressure shocks applied to fish eggs can be used to produce individuals that have three sets of chromosomes rather than the usual two. These triploid organisms generally do not channel energy into reproduction and thus are functionally sterile. Sex reversal can be accomplished by several methods including administering appropriate hormones. For example, genetically male tilapia can be turned into females through oestrogen treatments. These genetic males, when mated with normal males, produce a group of all-male tilapia.

Genetic Engineering in Livestock and Fish

Genetic engineering in animals can be used to introduce foreign genes into the animal genome or, alternatively, to "knock-out" selected genes. The method most used at present involves direct microinjection of DNA into the pronuclei of fertilised eggs, but progress is being made with new approaches such as nuclear transfer and the use of lentiviruses as DNA vectors. In the first genetic engineering experiments with farm animals, genes responsible for growth were introduced into pigs to increase growth and improve carcass quality.

Although conceptually simple, the methods used to genetically engineer livestock require special equipment and considerable dexterity, and no agricultural applications have proved commercially successful thus far. Applications in the near future therefore seem to be limited to the production of transgenic animals for use in the production of industrial or pharmaceutical products.

Genetic engineering is an active area of research and development in aquaculture. The large size and hardy nature of many fish eggs allow them to be manipulated easily and facilitate gene transfer by direct injection of a foreign gene or by electroporation, in which an electric field assists gene transfer. Gene transfer in fish has usually involved genes that produce growth hormone and has been shown to increase growth rates dramatically in carp, salmon, tilapia and other species.

In addition, a gene from the winter flounder that produces an antifreeze protein was put into salmon in the hope of extending the farming range of the fish. The gene did not produce enough of the protein to extend the salmon's range into colder waters, but it did allow the salmon to continue growing during cold months when non-transgenic salmon would not grow. These applications are still in the research and development stage, and no transgenic aquatic animals are currently available to the consumer.

Diagnostics and Epidemiology

Plant and animal diseases are difficult to diagnose because the signs may be misleading or even entirely absent until serious damage has occurred. Advanced biotechnology-based diagnostic tests make it possible to identify disease-causing agents and to monitor the impact of disease control programmes to a degree of precision not previously possible. Molecular epidemiology characterises pathogens (viruses, bacteria, parasites and fungi) by nucleotide sequencing, which enables their origin to be traced. This is particularly important for epidemic diseases, in which the possibility of pinpointing the source of infection can significantly contribute to improved disease control.

For example, the molecular analysis of rinderpest viruses has been vital for determining the lineages circulating in the world and instrumental in aiding the Global Rinderpest Eradication Programme (GREP). Enzyme-linked immunosorbent assay (ELISA) tests have become the standard methodology for the diagnosis and surveillance of many animal and fish diseases worldwide, and the polymerase chain reaction (PCR) technique is especially useful in diagnosing plant diseases and is proving increasingly so also for livestock and fish diseases. The effectiveness of plant and animal health programmes is also being considerably enhanced by the development of genetic probes that allow specific pathogens to be distinguished and detected in tissue, whole animals and even in water and soil samples.

Genetically Engineered Vaccines

Genetically engineered vaccines are being developed to protect fish and livestock against pathogens and parasites. Although vaccines developed using traditional approaches have had a major impact on the control of foot-and-mouth and tick-borne diseases, rinderpest and other diseases affecting livestock, recombinant vaccines can offer various advantages over conventional vaccines in terms of safety, specificity and stability. Importantly, such vaccines, coupled with the appropriate diagnostic test, allow the distinction between vaccinated and naturally infected animals. This is important in disease control programmes as it enables continued vaccination even when the shift from the control to the eradication stage is contemplated.

Today, quality improved vaccines are available for, for example, Newcastle disease, classical swine fever and rinderpest. In addition to the technical improvements, advances in biotechnology will make vaccine production cheaper, and therefore improve supply and availability for smallholders.

Animal Nutrition

Biotechnologies have already resulted in animal nutrition aids such as enzymes, probiotics, single-cell proteins and antibiotic feed additives that are already widely used in intensive production systems worldwide to improve the availability of nutrients from feeds and the productivity of livestock and aquaculture. Gene-based technologies are being increasingly employed to improve animal nutrition, either through modifying the feeds to make them more digestible or through modifying the digestive and metabolic systems of animals to enable them to make better use of the available feeds.

Although progress in the latter approach is likely to be slow because of gaps in our current understanding of the underlying genetics, physiology and biochemistry, one example of commercial success in high-input, intensively managed systems is the use of recombinant somatotropin, a hormone that results in increased milk production in dairy cows and accelerated growth and leaner carcasses in meat animals.

2

Biotechnology in Food and Agriculture

Unlike the Green Revolution, which was based on an explicit strategy of the international transfer of improved technology as a free public good, almost all transgenic crop varieties and most other agricultural biotechnology innovations are being created and disseminated by the private sector. However, several barriers stand in the way of biotechnology reaching resource-poor farmers and especially the poor countries that could benefit substantially from these innovations.

Safe and informed use of biotechnology requires adequate capacity for policy formulation, agricultural research, financial resources and marketing channels, as well as a framework for intellectual property rights and the capacity to handle the regulatory matters regarding food safety, human and livestock health, and environmental safety. Although biotechnology is evolving rapidly and is poised to play a fundamental role in further agricultural and economic development, there remains a large gap in most developing countries and especially among the least developed in their ability to assess their specific circumstances, meet their commitments and benefit from the opportunities that may arise from biotechnology. There is often a policy vacuum and inadequate capability to comply with the international instruments relating to biotechnology.

The most frequent problems encountered by developing countries and countries in transition are:

— insufficient capacity within ministries and their institutions to analyse options, set priorities for investment and formulate policies for national deployment of biotechnology in food and agriculture that support national development goals;

— limited technical, legal and administrative capability to establish and implement regulatory procedures, including biosafety, risk assessment and intellectual property rights, protection of indigenous knowledge and local resources, and communication to raise public acceptance of new technologies; and

— limited resources and capabilities to design, establish and operate infrastructures needed to generate, adapt, transfer and regulate biotechnology applications in food and agriculture, including enabling environments for furthering collaboration between the public and the private sectors.

Dynamic Capacities in Agricultural Biotechnology

Strong and dynamic capacity at the technical, institutional and management levels is the most important requisite for successful and sustainable application of biotechnology in food and agriculture. However, developing countries and countries with economies in transition vary widely in their capacity to manage agricultural biotechnology effectively. In particular, they span a wide spectrum in their capacities for agricultural biotechnology research and regulatory management, including intellectual property rights.

In recent years, there has been a steady development of agricultural biotechnology capacity in several of the larger countries, notably in Brazil, China and India, where human and financial resources allocated to biotechnology R&D are relatively high and experience in the commercialisation of biotechnology products is growing. Where governments have made substantial investments in public-sector agricultural biotechnology research in the developing world, several common themes emerge.

First, they have gradually built a strong scientific base in agricultural research and biotechnology. Their national research institutes are encouraged to be actively involved in bilateral and international collaborative research programmes in diverse fields of agricultural biotechnology. Second, in their national policies they have specifically identified science and technology, and biotechnology in particular, as an important engine of economic growth both for agriculture and for the health sector. Third, their public agricultural research programmes have had substantial success in promoting rapid agricultural growth. These countries have seen the explosive growth of information technology and its contributions to their economies and hope for similar growth through medical and agricultural biotechnology.

Towards the middle of the spectrum are those developing countries that are now beginning to incorporate biotechnology increasingly in their agricultural research programmes, for instance Egypt and Indonesia. These countries generally have moderately strong conventional agricultural research capacity and are developing strong biotechnology capacity in several areas. Further towards the other end of the spectrum are those countries that have not advanced far in direct application of the tools and techniques, except for applications of simpler techniques such as micropropagation and tissue culture. Again, these countries have several things in common.

Research efforts are less advanced and often several related programmes are scattered over a wide range of products and institutes. The programmes are often heavily dependent on donor funding and run the risk of drying up as soon as the funds are exhausted. Furthermore, the marketing and management of biotechnology products are virtually absent, as is the critical mass required to raise public awareness. In many instances, governments do not accord sufficient priority to agricultural research, and policies to support agricultural research in general and agricultural biotechnology in particular are either lacking or not implemented. Because advancements in agricultural biotechnology are severely constrained in these countries, potential pay-offs from biotechnology research and development programmes remain low.

The recently launched FAO-BioDeC is a database providing updated baseline information on the state-of-the-art biotechnology products and techniques that are in use or in the pipeline in developing countries and those with economies in transition. Currently, the database includes about 2 000 entries from 70 countries and focuses on research, testing and commercialisation of specific crop technologies and products in developing countries. Although the data are limited, they do give an overview of the different stages of adoption and development of these technologies in different countries and regions and offer the possibility of identifying gaps, as well as potential partners for joint programmes in areas of common interest.

In addition to research capacity, countries also vary widely in their capacity to regulate biotechnology. The spectrum ranges from those that have well-developed IPR regimes and food safety and environmental safety regulatory procedures to those that have little or no capacity to manage these issues.

International Capacity-Building Activities

A number of private, governmental, non-governmental and intergovernmental organisations are involved in one or more ways in capacity-building programmes in biotechnology. The focus areas include policy development assistance, research, technology transfer, biosafety measures and related regulatory oversight, development of associated legislation and creating public awareness. A wide range of activities are carried out for strengthening the policy, institutional and technical level of competence. Agencies involved in such initiatives are the International Service for National Agricultural Research (ISNAR) Biotechnology Service (IBS), the International Centre for Genetic Engineering and Biotechnology (ICGEB), the ISAAA, the Global Environment Facility (GEF), the United Nations Industrial Development Organisation (UNIDO), USAID and many more.

Although there is some overlap between the services offered by these organisations, each fulfils a certain function different to the others or places more emphasis on certain areas. There is no global information on the entire range of activities being carried out in agricultural biotechnology; however, the Biosafety Capacity Building database of the Biosafety Clearing-House provides a good overview of the various project activities being carried out in this area around the world.

FAO provides global fora to facilitate dialogue, and is a repository of statistical information. FAO can play a pivotal role in assisting Member Governments with science-based guidance on this subject as well as in standard-setting. Some of the key activities focused on biotechnology are as follows.

— Promoting international standard-setting bodies. FAO supports several agreements that have an important bearing on agricultural applications of biotechnology, particularly in relation to the WTO SPS and Technical Barriers to Trade (TBT) Agreements. These include the IPPC and the Codex Alimentarius Commission, and the International Treaty on Plant Genetic Resources for Food and Agriculture, which covers conservation and sustainable use of plant genetic resources for food and agriculture and the fair and equitable sharing of benefits derived from their use.

— Capacity building through technical assistance and training. FAO is assisting member countries in building their legal and regulatory frameworks in harmony with international obligations; training and strengthening facilities and institutions for the appropriate and safe utilisation of biotechnology for food and agriculture; developing their national legislations in this area; and building the capacity for participating in international negotiations on biotechnology to optimise national benefits.

— Information dissemination. FAO provides objective, science-based information on agricultural biotechnology, collecting, analysing and disseminating information in five languages, including through the corporate Web site and publications. This activity covers all aspects of biotechnology in food and agriculture, taking into consideration that member countries and their citizens need balanced and unbiased information on the potential benefits and risks of biotechnology.

In 2002, FAO and the United Nations Development Programme (UNDP) conducted an assessment of the status of biotechnology application in Bangladesh. Based on this assessment, the Government of Bangladesh formulated a National Programme for Biotechnology (NPB), which aims to utilise biotechnology as an important complementary route to fight food insecurity and poverty, two pressing problems of the nation. The NPB will promote awareness at all levels; establish and implement appropriate policies, strategies and partnerships; strengthen investment, institutional and market support; and undertake focused and integrated biotechnological research and development.

The key components of the NPB are:

— National Policy for Biotechnology, its implementation and governance. Address the technological and enabling aspects of biotechnology application. A National Taskforce for Sustainable Biotechnological Development (NTSBD), under the chairmanship of the Principal Secretary of the Office of the Prime Minister, will ensure that the policy is being effectively implemented. The NTSBD will provide transparent and efficient governance and build the required confidence in all stakeholders.

— Enabling regulatory measures. Legislative and regulatory frameworks for IPR, TRIPS, biosafety, and access to and nagotiations on new technologies and products, consistent with the national needs and farmers' aspirations and rights, will be established and strengthened. Effective containment facilities, risk analysis, other biosafety-related capacities and human resources to manage regulatory aspects have high priority. The introduction, evaluation and commercialisation of "Golden BR 29" (an elite Bangladeshi rice variety transformed at IRRI for high beta-carotene content) will be showcased to strengthen the national capacity in instituting and handling regulatory measures.

— Institutional strengthening. Biotechnological R&D institutions in the country will be strengthened by equipping them with state-of-the-art infrastructure, centralised facilities, suitably trained human resources, information and communication facilities and by fostering public-private partnerships. The capacity of NTSBD will be augmented for priority-setting, system-based decision-making, handling of issues in a disaggregated manner, cementing research-extension-farmer-market links and for generating and allocating resources.

— Biotechnology programmes. The NPB, following effective monitoring and evaluation, will focus on ecotechnologies towards an evergreen revolution, especially addressing the problems of small farmers. The following areas have been prioritised: production and distribution of in-vitro-cultured propagules, molecular characterisation of genetic resources, diagnostics and recombinant vaccine production, biocontrol of pests and diseases, production and commercialisation of quality (fish) fingerlings, development of transgenics for resistance to biotic and abiotic stresses, nutritional and other quality attributes, and molecular marker-assisted selection.

— Three developments to help Bangladesh realise its goal. (a) For the first time, Bangladesh has created a budget line for biotechnology in its national budget; (b) in order to ensure high efficiency and interdepartmental cooperation and to avoid wasteful duplication of effort, the NTSBD is being chaired by the Principal Secretary of the Office of the Prime Minister; and (c) UNDP and other donors and international organisations have shown considerable interest in funding the new initiative.

Challenges in Capacity Building Activities

Despite the range of capacity-building activities being carried out, much more needs to be done. The challenges faced are on a scale unlike those of other technological revolutions, including the Green Revolution in the 1960s and 1970s. For instance, any application of biotechnology requires a framework for safety including that of the environment and of human and animal health. There is a demand for equitable distribution of the benefits from the genetic resources utilised for biotechnology. In addition, it is important to develop consensus within society on the use of biotechnology-based products through full and transparent participation of all stakeholders in decision-making. Some of the major challenges in adoption of biotechnology include:

— ensuring resources to cover the high costs of inputs and development;

— building an enabling environment for the promotion of biotechnology;

— integrating biotechnology with conventional research programmes;

— addressing corporate control, market power and distributional implications;

— ensuring consumer protection and acceptance; and
— enhancing the sustainability of biotechnology programmes.

These factors, either directly or indirectly, affect capacity building, retention of personnel, and the balance between public- and private-sector capabilities. Although not exclusive to biotechnology, the initial costs of developing these technologies may increase the difficulties. Developing countries need to avoid the trap of dependence and unsustainability in their biotechnology programmes. Government policies should establish mechanisms to encourage both public- and private-sector investment and participation in agricultural biotechnology.

Public and private-sector research should be consciously complementary and not competitive. The policy framework should not only promote the safe use of biotechnology but also ensure that policies are not a deterrent to investment by the private sector and to collaboration with external partners. In many developing countries such progressive institutional and organisational reforms are hampered by the absence of appropriate policies or their appropriate implementation.

Recognising the constraints, there is a conscious need to take a sustained, holistic, multistakeholder, participatory approach to realise the potential benefits of agricultural biotechnology. In developing countries, there is a greater need to ensure not just capacity creation but also its retention and enhancement. Capacity-building activities have to be carried out at all levels: to raise the awareness of policy- and decision-makers, to initiate necessary legal and regulatory frameworks, to enhance technical and regulatory capacity, and to revamp institutions if necessary.

3

Characterisation and Conservation of Genetic Resources

United Nations Secretary-General Kofi Annan urged "individuals and institutions alike to give greater attention to biodiversity as a key theme in our efforts to fight the twin scourges of hunger and poverty and achieve the Millennium Development Goals". He also noted that the unprecedented loss of biodiversity over the past century was a major cause for alarm, where "many freshwater fish species, which can provide crucial dietary diversity to the poorest households, have become extinct, and many of the world's most important marine fisheries have been decimated.

Food supplies have also been made more vulnerable by our reliance on a very small number of species: just 30 crop species dominate food production and 90 per cent of our animal food supply comes from just 14 mammal and bird species—species which themselves rely on biodiversity for their productivity and survival. There has been a substantial reduction in crop genetic diversity in the field and many livestock breeds are threatened with extinction".

On the same occasion, FAO's Director-General Jacques Diouf also underlined that although forests are among the world's most important repositories of biological diversity, the world forest cover is decreasing at an alarming rate. Biotechnology is a broad collection of tools and these tools can be applied for a range of different purposes (e.g. genetic improvement of populations; disease diagnosis and vaccine development; improvement of feeds).

The biotechnology tools, such as molecular markers or cryopreservation and reproductive technologies, that can be used directly for the characterisation and/or conservation of genetic resources for food and agriculture.

Genetic Resources for Food and Agriculture

The domestication of crops and livestock, where species selected from the wild adapt to a special habitat created for them by humans, was a momentous development with major consequences for human societies, since "it provides most of our food today, it was prerequisite to the rise of civilisation, and it transformed global demography". Domestication seems to have arisen independently on at least five different occasions in different parts of the world where, for example, hunter-gatherers in the Fertile Crescent (an area in Asia, including parts of Iran, Iraq, Israel, Jordan, Lebanon, Syria and Turkey), began domesticating barley, peas, wheat, cows, goats, pigs and sheep about 10,000 years ago. Sometimes, the same species (e.g. cow, sheep) was domesticated independently in different places.

The process of domestication has continued ever since, e.g. the rabbit and strawberry were domesticated in the middle ages. Domestication had major impacts on human societies, making the transition from hunger-gathering to a settled farming existence possible. It also led to major changes in the species that were domesticated. Since their domestication, domestic crops and livestock have been used in almost all environments of the world and many have been used in particular environments for very long periods of time e.g. Soay sheep on Hirta, off West Scotland, are thought to be the survivors of the sheep first brought to Britain by the first farmers 5,000 years ago.

Each environment has a certain set of stressors that impacts the production environment. If we consider livestock, these might include climatic features (heat/cold, amount of rain, humidity); feeding conditions (quality/quantity vegetation available, access to water) or diseases (parasitic infections etc.) and, over time, populations adapted to the environment and its specific stressors. Farmers have also exercised a major influence on the domesticated species by preferentially selecting for specific morphological or production traits.

Unlike crops and livestock, domestication is generally a very recent phenomenon in the fishery and forestry sectors. Only a small number of forest tree species have been, to some degree, domesticated, and such action has been taken only over the past half century. Such tree species, used in plantation forestry, are usually no more than 1-2 generations removed from their "wild relatives".

For fish, apart from a few species, such as common carp domesticated about 2,000 years ago, aquaculture (fish farming) is a relatively new development and the majority of fish consumed today still come from wild populations, although it is predicted that by 2030 consumption from the ever-growing aquaculture industry will exceed capture fisheries. The agricultural genetic resources to be considered in this e-mail conference, therefore, include a wide range of species, from those domesticated 10,000 years ago to those domesticated in recent times and those which have not been domesticated but are nevertheless of importance for food and agriculture for humankind.

The need to conserve these resources is now widely accepted, generally justified for one or more of a variety of reasons, including their importance as an insurance against future changes in market needs and production conditions; as a source of material for scientific research and future germplasm development; and for cultural and historical reasons, because they are part of our heritage, passed down from previous generations.

The issue has received considerable attention at the international policy-making level. For example, it is addressed under the Convention on Biological Diversity, a legally-binding agreement with 188 Parties, through its agricultural biodiversity work programme, and under the International Treaty on Plant Genetic Resources for Food and Agriculture, a legally-binding treaty with 69 Parties that entered into force in June 2004.

Two major strategies for conservation can be distinguished: in-situ conservation, where a population is maintained in its natural or agricultural habitat, or ex-situ conservation, where it is maintained outside of this habitat. Ex-situ conservation can involve living individuals (e.g. animals in zoos, fish in aquaria, crops or trees in botanical gardens and arboreta) or tissues or genetic material from individuals (such as seeds, pollen, sperm, embryos, DNA). Ex-situ conservation is important for crops, where an estimated 6 million plant, mainly crop, accessions (i.e. samples of a variety collected at a specific location and time) are stored in national, regional and international collections.

Of the 6 million accessions, over 80% are in national collections and about 90% are stored in seed genebanks which, depending on the seed storage facilities, may allow short-, medium- or long-term storage. A genebank is a storage facility where germplasm of plant or animal origin is stored in forms such as seeds, pollen, semen or embryos; or in in vitro culture (germplasm kept as sterile plant tissues or plantlets on nutrient gels); in cryogenic storage; or in a field genebank i.e. as plants growing in the field. For the livestock, forestry and fishery sectors, in-situ conservation is currently more important than ex-situ conservation.

Characterisation of genetic resources goes hand in hand with their conservation as it is fundamental both for understanding what is being conserved and for choosing the genetic resources that should be conserved. Characterisation enables us to identify the key features (both the strengths and weaknesses) of the available genetic resources, and this knowledge can also be used to develop breeding programmes for sustainable use of the genetic resource to harness and/or disseminate the positive attributes identified in the population.

Characterisation can also play an important role regarding issues of access to, and benefit sharing from, agricultural genetic resources, as well the promotion and control of "bioprospecting" (i.e. biodiversity prospecting, the search for commercially valuable compounds, substances or genetic material in nature). If a country has characterised its genetic resources, it should be appropriately positioned to develop and implement conservation strategies for targeted species, and to promote and oversee any potential commercial or other enterprises arising from the knowledge generated; to protect the resources (or any of their products) from inappropriate bioprospecting, including that carried out by foreign companies or countries; and to negotiate appropriate compensation for use of these resources by third parties, be they national or foreign.

The kinds of features that could be characterised for the different agricultural populations include their morphological and physiological features; their phenotypic and economic performance in different environments; their interaction with, and relation to, the environment; their value as social or cultural objects; their role in scientific research; their diversity at the gene and DNA sequence level; their population size and degree of endangerment.

Crop Genetic Resources

Compared to the large numbers of wild plant species available, only very few were actually domesticated . Although this number is small, just a sub-set of them again are of special importance globally. For example, considering plant-derived energy intake worldwide, just three crops (wheat, rice and maize) provide over 50% of the intake and 30 crops provide 95%. Splitting the world into 17 sub-regions, there are only 12 major crops which supply more than 5% of the plant-derived energy intake in one or more sub-regions i.e. wheat, rice, maize, millet, sorghum, potato, sugar cane, soybean, sweet potato, cassava, the common bean and related species (Phaseolus), banana/plantain. It should, however, be kept in mind that although some specific crops may not have any global/regional significance they may, nevertheless be important at the individual country level (e.g. tef in Ethiopia).

Although, as seen, the number of major crops species is quite small, the diversity within each species can be very substantial. Within a species, cultivated crop varieties (cultivars) can be broadly categorised as farmer's varieties (also known as landraces or traditional varieties i.e. populations that are the product of breeding or selection carried out by farmers, either deliberately or not, continuously over many generations) and modern varieties (also known as high-yielding or high-response varieties, that are the products from professional breeders working in publicly-funded research institutes or private companies). Farmer's varieties tend to maintain high levels of genetic diversity and to be the focus of conservation efforts.

The first comprehensive worldwide report on the status and use of crop genetic resources was published by FAO in 1998. This State of the World's Plant Genetic Resources for Food and Agriculture (SoW-PGR) was developed through a participatory country-driven process and the primary data sources for the publication were detailed Country Reports from a total of 154 countries. The state of diversity within each of the 12 major crops is documented. It shows, for example, that the global ex-situ collection of millet contains roughly 90,000 accessions, of which 2% are wild relatives, 33% are landraces and old cultivars, 5% are advanced cultivars and breeding lines while 60% are unknown and mixed.

Although incomplete data makes it difficult to give a definitive picture of the kind of material stored in the world's ex-situ collections, it is estimated that roughly half the accessions for which the type of material is known are advanced cultivars and breeding lines, while a third are landraces and about 10% are wild species.

Forest Genetic Resources

Most forest tree species are characterised by inherently high levels of variation and extensive natural ranges. They are long-lived, frequently outcrossing organisms, with generally large distribution ranges. A high level of intra-specific genetic variation is needed to ensure present-day and future adaptability to changing environmental conditions. It is also needed to maintain options and potential for improvement to meet changing end-use requirements.

Forests provide a wide range of goods and services such as timber, fibre, fuelwood, food, fodder, gum, resins, medicines, pharmaceutical products and environmental stabilisation. Similar goods and services are often provided by a wide range of genera and tree species. Despite the availability of a large number of forest tree species, less than 500 have been systematically tested for their present-day utility for human beings and less than 40 species are included in intensive selection and breeding programmes.

Estimates for the year 2000 indicate that the world's forest cover was about 3.9 billion hectares, of which 47% was in the tropics, 33% in the boreal zone (i.e. the northernmost forest zone), 11% in temperate areas and 9% in the subtropics. About 95% of the forest cover was in natural forest and 5% in forest plantations (defined as forest stands established by planting or/and seeding in the process of afforestation or reforestation). Of the roughly 190 million hectares of forest plantation, 48% had an industrial end-use (e.g. pulpwood for paper); 26% a non-industrial purpose (e.g. fuelwood, soil and water conservation); and for 26% it was not specified.

Most of the forest genetic resources worldwide are therefore found in natural, largely un-managed forests which, however, are frequently intensively used for the provision of both wood and non-wood products, for local, national and international use. Conversion of forests to other land uses represents the greatest threat to forests and their diversity. According to FAO, "increasing pressure from human populations and aspirations for higher standards of living, without due concern to the sustainability of the resources underpinning such developments, heighten these concerns.

While some land use changes are inevitable, it is important that such changes be planned and managed to address complementary goals. Concerns for biological and genetic conservation should be major components of land use planning and forest management strategies".

Animal Genetic Resources

Of the 50,000 known bird and mammalian species, about 30 have been used extensively for agricultural purposes, with fewer than 14 accounting for over 90% of global livestock production. In the millennia following their domestication, animal species were influenced by both natural and artificial selection in a wide range of geographic and climatic areas and production systems, leading to the development of large numbers of genetically diverse domestic animal breeds.

Many of these breeds are currently endangered. In developing countries, the introduction of exotic breeds and their spread through indiscriminate crossbreeding is a major reason for loss of local, indigenous breeds. Other factors include changes in breeders' preferences because of short-term socio-economic influences; degradation of the ecosystem in which the breeds were developed; natural disasters (e.g. drought and diseases); wars and other forms of political unrest and instability.

In 1993, FAO published the first *World Watch List for Domestic Animal Diversity*, providing a basic global overview of the state of the genetic resources of seven mammalian species (buffalo, cattle, donkey, goat, horse, pig and sheep). The 3rd edition, expanded to cover 30 mammalian and bird species, showed e.g. that of the over

4,000 breeds with population data, roughly 30% could be classified at a high risk of loss and that chickens and cattle have large numbers of breeds at risk while the horse and goose have the highest percentages of breeds at risk of loss. In addition, apart from the wild boar and wild red jungle fowl (ancestors of the domestic pig and chicken respectively), the putative wild ancestors of the major livestock species are in danger of extinction or are already extinct.

In a similar process to that described for the crop sector, FAO is preparing the first State of the World's Animal Genetic Resources, that is expected to be completed by 2006. So far, 151 countries have accepted to submit country reports and the objective of the country and global assessments is to provide a comprehensive analysis of the status and trends of the world's farm animal biodiversity and of their underlying causes, as well as of local knowledge regarding its management.

Fishery Genetic Resources

Like forestry, and unlike crops and livestock, wild populations currently play a greater role than domesticated populations in the fishery sector. When considering the status of the world's fishery genetic resources, both the aquatic species that are fished (capture fishing) as well as those that are farmed (aquaculture) can be included. For capture fisheries in 2002, FAO Members reported that 974 taxa (i.e. taxonomic units such as a family, genus or species) of fin-fish, 143 taxa of crustacea, 114 taxa of molluscs, 26 taxa of plants and 73 taxa of miscellaneous animals (such as sea urchins, sea cucumbers and marine mammals) were taken from the world's capture fisheries. These numbers are, however, likely to be an underestimate as reporting is often incomplete.

Although over 1000 taxa are represented, about 10 species make up about one third of capture fisheries production. Unlike the crop and animal sectors, there is no systematic effort to describe the state of the world's fishery genetic resources below the species level. Nevertheless, description of the fish genetic stocks at the within-species level is important for fisheries management and is an active research area.

For aquaculture in 2002, FAO Members reported that, apart from a small number of aquatic plants and animals, a total of 153 species of fish, 60 species of molluscs and 44 species of crustaceans were farmed. The relative novelty of aquaculture can be shown by the fact that its contribution to global supplies of fish, crustaceans and molluscs has grown from just 4% in 1970 to 30% in 2002.

Aquatic science, in general, lags behind terrestrial sciences in identifying species, understanding ecosystem relationships and assessing potential uses for genetic resources. Communities of life on the ocean floor are poorly understood and, every year, some aquatic species become extinct before they have even been identified. Fish genetic resources worldwide are currently threatened by overfishing, but also by habitat destruction (e.g. dam building) and pollution from human activities.

Every 2 years, FAO publishes the State of World Fisheries and Aquaculture which, inter alia, provides a global overview of capture fisheries (marine and inland) and aquaculture. Regarding exploitation of marine fish stocks (estimated production was 84.5 million tonnes in 2002), the most recent edition indicates that there has been a consistent downward trend since 1974 in the proportion of stocks offering potential for expansion, coupled with an increase in the proportion of overexploited and depleted stocks, rising from about 10% in the mid-1970s to close to 25% in the early 2000s.

The percentage of stocks exploited at, or beyond, their maximum sustainable levels varies widely among fishing regions. The report says that the information available continues to confirm that, despite local differences, the global potential for marine capture fisheries has been reached, and more rigorous plans are needed to rebuild depleted stocks and prevent the decline of those being exploited at, or close to, their maximum potential.

For inland fish stocks (estimated production was 8.7 million tonnes in 2002), unlike the major marine fish stocks, they are less well defined and occur over much smaller geographical areas, such as individual lakes, rice fields or rivers, or over vast areas such as transboundary watersheds that are often situated in areas that are difficult to access. These factors make it costly to monitor the exploitation and status of fish stocks and, in fact, very few countries can afford to do so. As a result, the majority of countries report only a small fraction of their catch of inland fisheries by species, further compounding the problem of accurate assessment. There are, however, indications that these resources are undervalued and threatened by habitat alteration, degradation and unsustainable fishing activities.

Biotechnologies in Genetic Resources

A wide range of genetic resources are currently being used by humankind for food and agricultural purposes and many of them are in danger of being lost, due to reasons such as over-exploitation (fish), replacement of local with international germplasm (crops/livestock) or habitat change or destruction (crop/fish/forest/livestock).

Molecular Markers

Molecular markers usually do not have any biological

effect and are normally assumed to represent neutral loci unaffected by selection. They are identifiable DNA sequences, found at specific chromosomal locations on the genome and transmitted by the standard laws of inheritance from one generation to the next. A number of types of molecular markers exist, such as RFLPs, RAPDs, AFLPs, microsatellites and SNPs, which can differ in a variety of ways, such as the amount of time, money and labour needed or the amount of genetic variation found at each marker in a given population.

Molecular markers are used in a variety of approaches to characterise and conserve genetic resources, and here we will briefly describe the main ones. Some of the approaches are applied in each of the crop, forestry, animal and fishery sectors (e.g. estimating the genetic relationships between populations within a species) while others are more sector-specific (e.g. identifying duplicate accessions in crop genebanks or monitoring effective population sizes in capture fish populations).

Genetic Relationships between Populations

This is one of the major uses of molecular markers in agricultural research. Molecular markers are typed in populations of the same species that are e.g. located in different places and, by comparing the frequencies and presence/absence of marker loci in the different populations, inferences can be made about how genetically related they are to each other. The more differences in markers found between the populations, the greater the genetic distance between them and the populations are inferred to be less genetically related to each other than to populations that have more similar marker loci frequencies.

Taking a simple example for illustrative purposes (in practice many markers would be used simultaneously), if individuals in populations A, B and C are typed for a molecular marker M (with two alternative forms or alleles called M1 and M2) and all individuals typed in populations A and B have allele M1 while those in population C have M2, this suggests that A and B are genetically more similar to each other than to population C. In using molecular markers to estimate genetic distances between populations, it is assumed that they represent neutral loci (i.e. do not reflect genetic differences between populations at genes that are or have been under natural or artificial (farmer) selection) and that the use of a relatively small number of independently segregating marker loci will be a good predictor of the overall genomic diversity of a population.

Typically, a genetic diversity study involve sampling biological tissue (e.g. blood, leaves, fish scales) from a certain number of individuals from a number of populations of interest and typing them in the laboratory for a number of molecular marker loci. A statistical analysis of the marker data is then carried out to estimate the genetic relationships between the populations. For example, Lanteri and Barcaccia sampled 30 individuals each from 7 populations of wild cardoon identified in Sardinia and Sicily and each individual was genotyped for 32 microsatellite markers as well as 7 AFLP primer combinations. They used a statistical analysis method called principal coordinate analysis to show that the Sardinian and Sicilian populations were clearly differentiated, representing two distinct gene pools.

Such studies have been routinely applied in each of the agricultural sectors. For example, in developing a management strategy for commercially exploited fish, molecular markers are used to characterise the within-species genetic structure of the populations being harvested, to determine the units between which limited gene flow occurs i.e. to identify units which should not be overfished. A recent questionnaire-based survey of animal diversity studies carried out during the last 10 years provided information on a total of 86 projects, involving 13 livestock species from 93 countries. The majority of projects were for ruminants, especially cattle. Blood was the most frequently used biological sample and microsatellites were the most commonly used molecular markers.

As pointed out by Lenstra et al., these kinds of studies are not only relevant for the purposes of characterisation of genetic resources, they can also provide useful insights into the history of domestication. Evidence from molecular markers has been much used to elucidate whether individual crop or livestock species were domesticated just once at a single site or independently several times in different parts of the world.

Genebanks

Genebanks are an important tool for conservation of crop genetic resources, although currently of more limited importance for livestock, forestry or fish. Molecular marker information can be used in a number of genebank-related activities, such as sampling of material, management of the collections and stimulating use of germplasm stored therein. When sampling germplasm to create a genebank, information from the use of molecular markers to assess the within and between population genetic diversity can be used in conjunction with data on other characteristics of the populations to assist decision-making regarding which material should be sampled.

The amount of germplasm to be stored may be limited, due e.g. to financial reasons, so it would be important to sample material from genetically diverse individuals. Once a genebank has been established, molecular markers can also be used to assist in a number of genebank management activities.

As a first example, regeneration of seeds or other reproductive plant material in storage is an important task in a genebank, and the regeneration requirements (how frequently accessions need to be regenerated) depend on factors such as the species concerned, the storage conditions and the quality of the individual accessions. Markers can be used to check whether changes in alleles or in allele frequencies are taking place over time as a result of the adopted conservation strategies e.g. of frequent regenerations.

Secondly, while safety duplication of unique accessions acts as an insurance against possible accidents, unintentional duplication or overduplication of accessions is, however, wasteful and markers can be used to detect them.

Thirdly, to reduce costs and/or encourage greater and more efficient use by farmers and breeders, core collections (i.e. a subset of accessions selected to contain the maximum available variation in a small number of accessions) have been established for many crop species. Again, molecular markers can assist in this task.

Finally, molecular markers offer opportunities to increase the use of crop genetic resources from genebanks and they will therefore stimulate increased use of genebank material for research purposes, thereby playing an important role in the sustainable conservation of these genetic resources. Since a major justification for the existence of germplasm collections is use of the conserved accessions, De Vicente et al. underline the importance of genetically characterising them so that valuable genes can be identified for use in development of new plant varieties by conventional and advanced breeding methodologies.

Gene Flow

Wild relatives of domesticated species are of conservation interest as they represent one of the components of agricultural biodiversity. They have also proven useful for genetic improvement of a range of cultivated varieties. As described in the Background Document to Conference 7 of this Forum (on GMOs and gene flow), crossing of domesticated populations with their wild relatives has been well documented in some crop, forest tree, animal and fish species. For example, in crops, gene flow has been observed between rice and perennial rice, between maize and teosinte and between sugar beet and wild beet, while in animals there is evidence for crossing of domestic cattle with wild North American bison and of domestic pigs with European wild boars.

In forestry, gene flow from intensively bred forest trees to natural populations of the same or closely related species ("wild relatives") has been a cause for concern in many species, e.g. the European black poplar (Populus nigra). For fish, crossing of escaped farmed Atlantic salmon with wild Atlantic salmon is also a much-discussed problem. Gene flow from a domesticated population will cause changes in the genetic diversity of wild populations. In cases where diversity is systematically reduced over time by such geneflow, it may potentially lead to genetic extinction of the wild population in its original genetic state. Molecular markers can assist in conservation of the wild relatives through their use to distinguish hybrids from non-hybrid wild relatives and to monitor gene flow from domesticated populations to their wild relatives.

Effective Population Size

When developing conservation strategies for wild species, accurate estimates of the effective population size (Ne) are important because Ne predicts a number of parameters such as the rate of inbreeding, and increases in inbreeding (due to mating of genetically related individuals) can lead to the reduced abilities of individuals to survive and reproduce (a phenomenon called inbreeding depression). Ne can be estimated using pedigree information and/or knowledge about the population breeding structure (male/female mating ratio, family sizes etc.). However, for wild populations, especially of aquatic species, it can be difficult to get this information.

Molecular markers can be used to estimate Ne, something which has been done, for example, in several commercial fish populations. Similarly, molecular markers can also be used as a tool to monitor fish population sizes over time and to detect any major recent declines in Ne.

Cryopreservation

There are two major strategies for conservation, in-situ and ex-situ, the latter involving e.g. storage of genetic material in genebanks or keeping live animals or plants in zoos or botanical gardens respectively. Biotechnology, through cryopreservation (freezing) technologies and in vitro culture, provides additional tools for ex-situ conservation. Cryopreservation involves the preservation of germplasm in a dormant state by storage at ultra-low temperatures, usually in liquid nitrogen (-196C) and it can be used to preserve biological material (e.g. seeds, sperm, embryos) of crop, livestock, forest or fish populations indefinitely in genebanks. It allows large quantities of genetic material to be stored in a limited amount of space.

If the living populations later become extinct, they can in many instances be regenerated using the frozen genetic material. In addition, if the living populations are declining in numbers and develop problems due to inbreeding, these problems can be alleviated by introducing unrelated genetic material from the genebank. In animals, collection of germplasm for cryopreservation and production of live offspring from cryopreserved genetic material depends on a number of associated reproductive technologies, such as embryo transfer.

In vitro culture is a plant biotechnology that involves application of slow growth procedures to germplasm accessions kept as sterile plant tissues or plantlets on nutrient gels. Information provided on cryopreservation in livestock/fish and crops/forest trees is based primarily on Hiemstra et al. and Panis and Lambardi, and the references therein, respectively.

Livestock and Fish

In sexual reproduction, through fertilisation there is a fusion of 2 haploid gametes to form the diploid zygote, a cell which begins dividing and develops to become an embryo (note, haploid and diploid cells have one or two sets of chromosomes respectively). The female gamete is called the ovum or egg, and is derived from a cell called an oocyte that has undergone two meiotic divisions. The male gamete is called the sperm (contained in fluid called semen).

For cryopreservation purposes, semen is currently the biological material of greatest importance, although embryos and, to a lesser degree, oocytes and somatic cells, have also received considerable research attention. Success in freezing these materials varies considerably from species to species, which is not surprising given the wide diversity in reproductive systems (e.g. fertilisation is usually external in fish and internal in birds, where the female stores the sperm before fertilisation). Collection of biological materials for the genebank and production of live offspring following thawing of formerly cryopreserved material is made possible through a variety of reproductive technologies, such as artificial insemination (AI) using semen, multiple ovulation and embryo transfer using embryos etc.

Freezing of semen and subsequent successful AI with thawed semen was first achieved over 50 years ago and semen of most livestock species can now be frozen adequately. There are large differences between species in insemination techniques and pregnancy rates using fresh or frozen semen, where AI with frozen semen is most successful in cattle. For domesticated birds, although AI is less widely used than in mammals, reasonable success rates (albeit at times highly variable) with frozen semen have been reported for the major bird species.

For fish, sperm cryopreservation has been tested in over 200 fish species with external fertilisation and "the present state of the art for many species of fish seems to be adequate for the purpose of gene banking". For embryos, cryopreservation in cattle is now a routine procedure. It has also been reported for other mammalian species, such as horse, pig, rabbit and sheep.

Embryos to be cryopreserved can be collected using non-surgical (cattle and horses) or surgical methods (other mammalian species). Inducing multiple ovulation using hormones can increase efficiency of the process. Alternatively, embryos can be collected through in vitro maturation and in vitro fertilisation of oocytes (a more technically demanding approach). Production of live offspring following cryopreservation involves non-surgical (only in cattle and horses and, to a lesser extent, pigs) or surgical embryo transfer to recipient animals.

For bird and fish species, however, cryopreservation of embryos has not been successful, mainly because of the large size, the high lipid content and the polar organisation of the ova and early embryos. For mammalian oocytes, live offspring have been reported in cattle and horse from embryos produced using cryopreserved oocytes, although the efficiency and reliability of using oocytes for generating offspring is still much lower today than with cryopreserved embryos.

Oocytes for cryopreservation in the genebank can be collected from ovaries of dead animals or by ovum pick-up (recovery of oocytes using an ultrasound probe) on live animals. Production of live offspring following thawing of cryopreserved oocytes involves in vitro maturation and fertilisation and eventual embryo transfer to recipient animals. Cryopreservation of bird and fish oocytes has not yet been successful, mainly for the same reasons given earlier for embryos.

For somatic cells (i.e. cells not involved in sexual reproduction), a number of cell types (e.g. skin fibroblasts, mammary cells) can now be cryopreserved. Collection of material for the genebank is easy and cheap. Production of live offspring (which, apart from mitochondrial DNA, are genetic clones of the individual which donated the somatic cells) is, however, complicated and has a low rate of success. It involves culturing the thawed cells and then transferring their nuclei (i.e. the organelle which contains the genetic material) to, or fusion of the somatic cells with, enucleated (i.e. without a nucleus) oocytes from another individual.

The resulting embryos are then cultured and eventually transferred to recipient animals. In mammals, live offspring have been obtained from embryos generated from somatic cells in a number of mammalian species (cattle, goats, horse, pigs, rabbits and sheep) and in fish (zebrafish), but not yet in poultry. The low success rates, as well as the many developmental problems encountered, are current major limitations to application of the technology.

Crops and Forest Trees

About 90% of the 6 million plant accessions in genebanks, mainly crops, are stored in seed genebanks. However, storage of seeds is not an option for crops or trees that do not produce seed (e.g. bananas) or that produce recalcitrant or non-orthodox seed (i.e. seed that does not survive under cold storage and/or the drying conditions used in conventional ex situ conservation, e.g. mango, coffee, oak and several tropical forest tree species).

Likewise, for plant species that are propagated vegetatively to preserve the unique genomic constitution of cultivars, such as fruit and several timber and ornamental trees. It also presents practical problems in most forest tree species in regard to the necessary, periodic regeneration of seedlots, as tree species usually have a long vegetative period prior to producing flowers and seed (lasting from several years to several decades), and as they require large areas to carry out such an operation, due to outcrossing and their size.

Genetic resources of these species are normally conserved either in vivo, i.e. living collections, or by in vitro culture, where slow growth procedures allow the plant material to be held for 1-15 years under tissue culture conditions with periodic regeneration (subculturing), depending on the species. Normally, growth is limited using low temperatures often in combination with low light intensity or even darkness.

Temperatures in the range of 0-5C are employed for cold tolerant species and 15-20C for tropical species. Growth can also be limited by modifying the culture medium and reduction of oxygen levels available to the cultures. The former option (in vivo) is the cheapest and thus the most widely adopted, but exposes the material to the harms of the external environment, e.g. contamination with pathogens or parasites.

In vitro culture, on the other hand, ensures that the collections are kept free from pest and diseases and protected from adverse conditions. Maintenance of in vitro collections is, however, labour-intensive and there is still a risk of losing accessions due to contamination, human error or somaclonal variation, i.e. mutations that occur spontaneously in tissue culture, with a frequency that increases with repeated subculturing.

In all these situations, as well as for long-term storage of seeds from orthodox species, cryopreservation offers an alternative tool or complementary strategy for ex-situ conservation. Following plant cell, tissue or organ storage at low temperatures, plants can then be regenerated and there seems to be no clear evidence so far of morphological, cytological or genetic alterations arising as a result of cryopreservation.

They, however, conclude that, despite the fact that cryogenic procedures are now being developed for an increasing number of recalcitrant seeds and a wide range of tissues and organs, the routine utilisation of cryopreservation for the preservation of plant biodiversity is still limited. They note that the main drawback for a wider application of plant cryopreservation is the unavailability of efficient cryopreservation protocols for many plant species. They report, nevertheless, that for various herbaceous (i.e. non-woody plants), hardwood (i.e. broadleaf, deciduous trees) and softwood species (i.e. coniferous trees), cryopreservation of a wide range of tissues and organs such as cell suspensions, embryogenic cultures, pollen, meristematic tissues and seeds has been achieved.

Cell suspensions (cells in culture in moving or shaking liquid medium) have been cryopreserved from banana and grape while cryopreservation of embryogenic cultures (i.e. callus or suspension cultures with potential to differentiate somatic embryos) is an advanced technology in conifers and is already being successfully applied to numerous commercial species, including pines. Methods exist for the cryopreservation of pollen from many crops, but its application is still rather limited and restricted to a few research centres.

Meristematic tissues (also called meristems, i.e. plant tissues, such as shoot tips, in which the cells are capable of active division and differentiation into specialised tissues such as shoots and roots) are the most common explants (i.e. a portion of a plant aseptically excised and prepared for culture in a nutrient medium) for the cryopreservation of vegetatively propagated species, such as fruit trees and many root and tuber crops.

There is large scale application of shoot tip cryopreservation in fruit crop germplasm collections, such as in plum and apple, while the number of cryopreserved herbaceous meristem samples is significantly smaller, although continuously growing. Seeds of most common agricultural and horticultural species can be cryopreserved and for orthodox seeds this can be considered as an alternative, as it offers long-term storage, to traditional storage (e.g. in celery).

4

Conservation of Plant Genetic Resources

In the field of plant genetic resources for food and agriculture (PGR) the irreversible loss of single genes or combinations of genes in genotypes, the so-called gene-erosion, is of major concern. Such genetic diversity is the primary source for variation in morphological and physiological appearance of plants. Diversity is the basic factor of evolution in species. It made it possible for crops to be adopted to the most different environments and uses, and genetic diversity will allow them to respond to the challenges of the next century.

Plant genetic resources for food and agriculture are a reservoir of genetic adaptability, which acts as a buffer against harmful environmental changes and economic challenges. The erosion of these resources results in a severe threat to the world's long term food security. Although often neglected, the urgent need to conserve and utilise plant genetic resources as a safeguard against an unpredictable future is evident. The complexity of serious ecological problems, partly also caused by agricultural science, now to be solved with the help of agricultural sciences, is a great challenge of the "century of environment".

The disciplines of soil sciences and the search for renewable resources from plants as raw materials are important issues, which are related to research in plant genetic resources. The future development of agriculture must become much more aware of ecological and environmental impacts. External effects, which have so far hardly been part of economically based decisions, have to be taken into consideration. This requires renewed approaches, for the management of and research in biodiversity. These changes are as substantial as other changes in biological sciences, which can be observed today, and can be compared to what is called a paradigm shift. It will be necessary to better understand the threats to plant genetic resources, including the consequences of introduction of new species, genetically modified organisms, change in agricultural practices and land use, and pollution of agroecosystems.

A substantial change in approach or even a paradigm shift regarding conservation of agrobiodiversity and ecosystem-management from ex situ conservation to on-farm resp. in situ conservation can be observed. A similar change is ongoing regarding the strategies for most effective study and utilisation of ex situ collections of plant genetic resources. The impact of modern biotechnology on agriculture, plant breeding and, thus, agroecosystems also causes severe changes. The severe changes mentioned require for an effective early warning system on plant genetic resources.

Plant genetic resources (PGR) form a distinct group in the entire concept of biodiversity. They are made up by the species of cultivated plants and their wild relatives. From the viewpoint of a specific crop, PGR belong, according to Harlan and de Wet, to the primary and secondary genepool of the respective species. In general PGR include all species, which contribute to people livelihoods by providing food, medicine, feed for domestic animals, fibre, clothing, shelter and energy and other uses, and are subject to cultivation by man as well as the wild relatives of these species. However, in comparison to the total number of species of higher plants which have been identified world-wide, PGR comprise 40 % of these species, while the crop plants themselves cover only 2.8 % of the species (Table 1).

Table 1. Number and relative amount of botanical species of higher plants, plant genetic resources (PGR), and crop plants in Germany, Europe, and world wide

	Higher Plants	PGR	Crop Plants
Germany	2,500 (100 %)	1,055 (42.2 %)	150 (6.0 %)
Europe	11,500 (100 %)	4,730 (41.1 %)	500 (4.3 %)
World	250,000 (100 %)	100,000 (40.0 %)	7,000 (2.8 %)

It is sometimes stated that only the 30 crop species "feed the world". The most important of them are listed in Table 2. These species provide more than 90 % of calories or protein to human nutrition. Therefore, improvement by modern plant breeding and conservation of their diversity has always received attention.

However, the number of species used for peoples livelihoods is much larger than these thirty and is estimated to be close to 7,000 botanical species. The amount of PGR accessions in ex situ collections world-wide is estimated to be 6.1 million accessions (Table 3). The 30 species listed in Table 2 already cover 4,036,000 accessions, or 2/3 of all genetic resources in ex situ collections. This is an important point to have in mind when discussing plant genetic resources for food and agriculture. Several crop species are neglected by agricultural science or breeding efforts, and their diversity and genetic resources do not receive the necessary attention to protect the loss of important diversity.

Table 2. The thirty crops with the highest number of accessions represented in ex situ collections

Crop	Total accessions world-wide	Crop	Total accessions
Wheat	784,500	Chickpea (Cicer)	67,500
Barley	485,000	Prunus	64,500
Rice	420,500	Clover	61,500
Maize	277,000	Capsicum	53,500
Garden Bean	268,500	Cotton	49,000
Oat	222,500	Grape	47,000
Soybean	174,500	Triticale	40,000
Sorghum	168,500	Medicago	33,000
Brassica	109,000	Sweet potato	32,000
Apple	97,500	Potato	31,000
Millet (Panicum)	90,500	Faba bean	29,500
Cowpea (Vigna)	85,500	Sunflower	29,500
Groundnut (Arachis)	81,000	Lupin	28,500
Tomato	78,000	Cassawa	28,000
Pea	72,000	Rye	27,000

Weeds, which have been subjected to co-evolution with cultivated plants in agroecosystems, developed also a wide range of diversity and sometimes have potential value for cultivation, are hardly collected systematically. There is a lack of knowledge about the diversity and geographical distribution of less utilised agriculture species. The need for more attention to such crops in conservation and utilisation programmes is evident. The knowledge about the species cultivated world-wide for food and agriculture is still increasing. Therefore, the estimated number of species of cultivated plants has increased considerablv

Methods for Agrobiodiversity Conservation

The world-wide existing ex situ collections of PGR contain approximately six million accessions (Table 3). However, there is duplication within and between collections. The total number of unique accessions in collections, which have been established to preserve diversity, is estimated to be one or two million only. 40 % of all accessions in genebanks are cereals. Food legumes comprise about 15 %, vegetables, roots and tubers, fruits, and forages account each for less than 10 % of the global collections. Spices and medicinal, aromatic and ornamental species are rarely found in genebanks. More than 1 million accessions belong to the three main crops wheat, maize and rice.

Table 3. Number of ex situ collection and conserved accessions, by region.

Region	Genebanks		Accessions	
	Number	%	Number	%
Africa	124	10	353.523	6
Asia	293	22	1.533.979	28
Europe	496	38	1.934.574	35
Near East	67	5	327.963	6
North America	101	8	762.061	14
Latin America and Caribbean	227	17	642.405	12
Total	1.308	100	5.554.505	100
CGIAR			593.191	
Total			6.147.696	

About half of all ex situ accessions are advanced cultivars or breeders lines, while just over one third are landraces or old cultivars, and about 15 % are wild or weedy plants including crop relatives. Characterisation data exists on only one third of these accessions. There is a gap in the collections regarding minor crops and underutilised species. In particular landraces and wild relatives of cultivated plants from their centres of diversity and cultivation are not represented adequately. Exploration of selected minor and underutilised species for germplasm collections and assessment of genetic diversity of landraces, therefore, has priority in the field of crop plant research.

The exploration, collection, conservation and identification of potentially valuable plant genetic resources for food and sustainable agriculture, which are endangered by extinction, as well as other plant genetic resources, which have potential value for future development, are primary obligations for all countries and institutions adhering to the FAO International Undertaking on Plant Genetic Resources. The Global Plan of Action for the conservation and sustainable utilisation of Plant Genetic Resources for Food and Agriculture emphasises, besides the in situ and on-farm conservation, the importance of ex situ conservation.

The figures presented in Table 3 show that the total number of accessions in ex situ collections is very large. But, besides the mentioned problem of duplication, the security of ex situ conservation is also questionable. About half of all genebank accessions world-wide urgently require rejuvenation, and in several countries this amount is even larger. In Albania 80 % of the accessions conserved ex situ require regeneration, which is alarming, since gene erosion in agroecosystems in Albania is still observed, while other industrialised countries already lost most of this on-farm diversity during industrialisation. Lack of funds is a general complaint of genebanks, and political instability can also threaten ex situ collections.

Although there are high numbers of accessions for some important crop species in the collections, it is difficult to estimate, which proportion of the entire diversity of such a species is really preserved in ex situ collections. Under on-farm conditions the diversity is always subjected to evolutionary changes and adaptations, while ex situ collections usually try to do exclude genetic shift and drift during rejuvenation and long term conservation of germplasm. The continuing research in biodiversity has resulted in definition of different categories of biological diversity. These are:

1) genetic (infraspecific) diversity,
2) diversity of species, and
3) diversity of ecosystems.

There exist strong interactions between these categories of diversity. From the viewpoint of plant genetic diversity for food and agriculture it has within each category to be differentiated between diversity of (1) cultivated plants, (2) wild relatives of cultivated plants, and (3) weeds. It is possible to analyse, which strategy for conservation of biodiversity is the most appropriate one for each of the mentioned categories and for each specific group within these categories. The conservation strategies can be categorised into: (1) ex situ conservation (genebanks, botanical gardens), (2) on-farm conservation (agroecosystems), and (3) in situ conservation (natural ecosystems).

Table 5 scores the advantages of each conservation strategy according to the respective category of biological diversity and the specific groups within these. Complementary action between these strategies is necessary. This needs increased cooperation in times of general reduction of budgets for such work, which can be more effective, if the relative superiority of the different conservation methods are taken into consideration.

Ecogeographic surveys of the ranges of distribution of a species as they have been initiated by N. I. Vavilov, are still the basic tool for developing adequate strategies for monitoring and collection of relevant diversity of a plant species. Such studies deliver basic information about: (1) the geographic distribution of a particular species, (2) patterns of infraspecific diversity, and (3) the relationship between survival and frequency of variants and associated (agro-)ecological conditions.

In cultivated plant species such studies are essential to understand the process of domestication and diversification on the infraspecific level. The identification of primary and secondary centres of diversity is based on such studies, which are still important. As an example in this direction Pistrick presented the range of distribution of wild radish, Raphanus raphanistrum L. ssp. raphanistrum. He could show for the same species, how the historical exchange of crop genetic diversity over far distances has led to the development of secondary centres of diversity for this subspecies.

A number of methods are available to analyse genetic diversity. Since each of these provides different types of information, the choice of the appropriate method or methods depends upon the information required as well as the resources and technological infrastructure available. Diversity can be studied, measured and quantified at different organisational levels, which are subject to research on ecosystem, plant sociology, species, infraspecific, cellular, sub-cellular, or molecular levels. Whatever the level of analysis may be, the purpose is to detect and quantify diversity. The basic questions are: What are the suitable units for quantifying diversity at the respective levels? What kind of variation is of importance for the question to be investigated?

Table 4. Conservation methods and their relative superiority for different categories of diversity scored by their importance for specific groups of diversity

Conservation method	Ex situ (Genebanks)	On-farm (Agroecosystems)	In situ (Other ecosystems)
Diversity-category			
Infraspecific Diversity	C***	C**	C^{0}
	R*	R^{0}	R***
	W*	W**	W*
Diversity of species	C**	C***	C^{0}
	R*	R^{0}	R***
	W*	W**	W*
Diversity of ecosystems	C^{0}	C**	C^{0}
	R^{0}	R^{0}	R***
	W^{0}	W**	W^{0}

Explanations: The relative importance of the methods to the specific groups of diversity is indicated by the number of stars
C= Crop species
R= Species of wild relatives of crops
W= Weeds
* = low importance
** = important
***= very important
0 = no importance

For research in cultivated plants, for handling and utilisation of diversity of plant genetic resources for food and agriculture, measurements of the extent of variation, i.e. the polymorphism, can easily be made by observing the phenotypes of plants. If such characterisation uses clear-cut qualitative traits, like colour, morphology or enzyme variants, it already reflects to high degree the genetic diversity by being relatively independent from environmental influences. More technical methods are needed to analyse genetically complex quantitative traits that vary continuously over a scale. These included agronomic traits such as yield or plant height, which are more typical for evaluation data in the PGR context.

Morphological and agronomic characters are often used for basic characterisation, because this information is of high interest for users of genetic diversity of PGR. Such characterisation requires human labour and organisational skills as well as an elaborated system of data documentation. It can be done using simple techniques and is capable to reach a high sample throughput. The data is not only of interest for users of the collection, but allows interpretation of the relations between the genotype and the environmental conditions, and, thus, conclusions on the evolutionary responds of the given species to selection pressures. This is essential for understanding the evolutionary pathways of cultivated plants.

Taxonomical treatments of infraspecific variation tend to favour qualitative traits in the analysis of diversity. As basic units for measurement of diversity, taxonomy defines infraspecific taxa like subspecies and botanical varieties. Breeders, however, to a large extent have to focus on quantitative traits in reaching their breeding goals. Quantitative agronomic traits measure the differences between individuals and populations with regard to genetically complex traits such as yield potential and stress tolerance. The diversity of a population with regard to such a complex trait can be described by using its mean value and genetic variance in statistical terms.

The traits detected are of high interest but frequently subject to strong environmental influence, which makes their use for defining units for measurement of genetic diversity difficult. Taxonomical classifications are not based on such characters, while informal classifications, which are sometimes used in agriculture, are frequently based on such traits. The relationship between the curators and users of PGR is a key issue in the management of ex situ collections. The analysis of genetic diversity is necessary to monitor genetic erosion, to guide collecting priorities, to investigate the evolutionary history of crops, to support breeding strategies, to guide the management of germplasm collections, to recognise genetic erosion, and to established an early warning system.

The major advantage of molecular methods for characterisation is their direct investigation of the genotypic situation, which allows them to detect variation at the DNA level and thereby excluding all environmental influences. Furthermore, characterisation can usually be done at early growth stages. The advantages and disadvantages of some commonly used molecular techniques for characterisation of PGR are summarised in Table 5. According to the chosen method, they can be very sensitive to any genetic differences and, thus, detect much more genetic diversity than the classical morphological methods.

Molecular marker techniques have become powerful and accurate tools for the analysis of genetic diversity. If molecular markers are correlated to characters of interest in breeding programmes, they are invaluable and help to accelerate breeding progress considerably. Also for quantitative traits of interest in plant breeding molecular markers (quantitative trait loci) have been found in some cases. Tanksley & McCouch stated: "New findings from genome research indicate that there is tremendous genetic potential locked up in seed banks that can be released only by shifting the paradigm from searching for phenotypes to searching for superior genes with the aid of molecular linkage maps". The tendency to use molecular marker techniques to assess diversity has already marginalised other methods used for this purpose.

Nevertheless, markers useful to characterise the plant genetic diversity and in particular PGR are not only molecular ones. It is important to identify as many markers as possible of different nature, which support analysing and quantification of diversity and genetic erosion. This will help to develop scientifically improved units for measurement of genetic diversity. The classical infraspecific taxa have been a first approach for creating an overview by establishing systems, and these systems are open for addition of further information, which results from molecular research. A broad approach is required to analyse diversity and to support conservation, management and development of plant genetic resources. Better methods for quantification of biological diversity are essential in order to describe diversity and to get a clear picture of the ongoing gene-erosion.

5

Genetic Improvement

Data Processing for Genetic Improvement

Data processing for genetic improvement requires a considerable amount of computation. The methods used depend on the nature of the breeding programme and the information required before selection decisions can be made.

Environmental Effects

Environmental factors tend to obscure true genetic ability. Performance records should therefore be adjusted for known environmental effects to remove some of the environmental variation. The value of such adjustment depends on the magnitude of the environmental effects and the accuracy with which they can be estimated. The procedures to correct for environmental effects are follows.

The simplest procedure, which is also efficient, is to relate the individual's record to the average of that of other animals subject to similar influences. Considering sheep for example, the weaning weight of a male lamb born as single to a 2-year-old ewe should be compared with the mean of other animals in the same category. This is possible only in large populations. In small populations, the method suffers from high sampling errors. Correction factors are therefore commonly used to adjust for environmental effects.

Correction factors

In principle, correction factors are only valid for the populations from which they are estimated. Estimates made within herds are, however, imprecise because of the small amounts of data. Furthermore, the computational work involved is also heavy. In practice, therefore, correction factors are estimated from data of many herds in a region and these are then used in all herds in that environment. Correction factors estimated from simple averages of group means are liable to be biased because of disproportionate numbers of these groups among other environmental classes. For example, correction factors for type of birth in sheep obtained from simple averages of single and twin-born lambs will be biased because of unequal numbers of these two groups in other classes such as dam-age or sex of lamb. Calculation of unbiased estimates requires simultaneous consideration of all effects. This is done by the least squares procedure.

The least squares procedure disentangles the different kinds of environmental effects. Before data can be prepared for least squares analysis by computer, the various environmental factors to be considered should be included in a model. The model can include continuous variables, e.g. age at weaning, and discrete classes e.g. sex. Interactions, if appropriate, may be included but they increase the amount of computation. It may be possible on the basis of preliminary analysis to omit some of the classes.

There should be at least two classes for a least squares analysis. Before assigning subclasses within a class, an estimate of the number of animals in each subclass should be made and where possible small subclasses should be combined or redistributed if the procedure is biologically acceptable. The analysis is more efficient if there are fewer subclasses and more individuals per class.

Correction factors are used in two ways to adjust data. The first adjusts all observations to the fitted mean by subtracting the constant from the observed value. The other adjusts all observations in relation to a subclass which is chosen as the standard (needs no correction), e.g. in adjusting for weaning weight of beef calves, the weaning weight of male calves born to dams over 6 years of age may be chosen as the standard. The latter method is more realistic as the adjusted data can be compared to those from other sources.

Environmental effects can also be corrected for by using a combination of correction factors and within subclass deviations. In this procedure, the data are first adjusted for all effects except those of herd and year using standard correction factors. Adjustments for the latter are then made by expressing the adjusted records as within-herd-year deviations. In small herds, the animal's own record should be excluded in computing the herd average, as otherwise autocorrelation arises between the herd mean and the animal's own production.

External and internal factors

The performance of an animal can be influenced by external and internal factors. External factors affect the whole herd or flock, e.g. herd, year, season, while internal factors affect the individual animal, e.g. weaning age.

Dairy cattle

Herd, season and year differences: The usual procedure used to remove herd, year and seasonal variations are the contemporary comparison and herdmate comparison procedures. Both procedures are based on deviations of the individual performance from the average performance of animals milking in the same herd at the same time. But, while the contemporary comparison uses only the first lactation records, the herdmate method uses age-corrected records of all the animals commencing lactation during the year and season under consideration.

When the number of first lactation animals in a herd within a year and season is small, contemporary comparisons are inefficient. The herdmate method is therefore to be preferred when herd sizes are small. Different procedures are used to obtain herdmate averages. Sweden uses the rolling herd average of which is obtained by dividing the total milk produced in the last 12 months by the total number of cow-years. In Denmark, it is calculated as the average of all corrected 305-day lactation yields completed during the past 12 months.

Settled village dairy herds in the tropics are usually very small in size. In such situations, the concept of a herd as a unit may have to be abandoned and herds in a location grouped into units on the basis of similarity of breed, production level and management.

Age at calving: Adjustment for age effects on first lactation performance is done on the basis of the regression of yield on age. Lactation yields of older cows are adjusted on the basis of parity to either first lactation or mature equivalent basis. Another method of correcting such records is on age, ignoring parity. This may be suitable when year and parity effects are confounded as when a new herd is started with young cows.

Lactation length: The 305-day lactation yield represents the yield from calving to 305 days irrespective of the length of lactation. when cows of European breeds are dried off before 305 days for management reasons, short records are corrected to 305 days. Such procedures are, however, not valid for cattle in the tropics, where short lactations appear to have a genetic basis. Calf death can also result in short lactations in zebu breeds that have a strong maternal instinct. However, correction for calf mortality may result in an upward bias in animals that also have a low potential for lactation length.

Calving interval: Milk yield is influenced by both the preceding calving interval and by the number of days pregnant during the current lactation. Since the relationship between length of calving interval and milk yield is curvilinear, regression using the linear and quadratic terms should be used to estimate the correction factors.

Beef cattle

Correction factors for weaning and post-weaning weights are commonly used in developed countries. The Meat and Livestock Commission in UK uses a system that combines correction factors and within-subclass deviations. The correction factors adjust only for age of dam. In USA, however, standard correction factors are used to adjust for all the effects. The UK system avoids the errors arising from the use of standard correction factors but suffers from the difficulties associated with inadequate number of contemporaries in small herds. Standard correction factors are, however, not available in most tropical countries. The following environmental factors should be considered in their computation.

a. *Herd and year*: These effects are similar to those in dairy cattle.

b. *Season*: Variations in forage availability caused by marked differences in rainfall during the dry and wet seasons have a profound influence on performance of beef cattle in the tropics. Weaning weight depends on the season of birth while post-weaning weight is related to the season in which an animal reaches a given age.

c. *Age of dam*: The effect of age of dam on weaning weights in zebu cattle in the tropics may be less pronounced than in temperate countries because of the late age at first calving.

d. *Previous parous state*: Calves raised by dams parous the previous year are generally lighter at weaning than calves from dams which were non-parous and thus rested the previous year. The classes into which animals should be grouped are parous, parous but not having raised a calf and non-parous.

e. *Current calving interval*: The pregnancy status of the cow during weaning can influence weaning weight. However, it is not practicable to adjust for this effect because of the difficulty in obtaining mating records.

f. *Age at weaning or post-weaning*: The usual procedure for adjusting for age variations is to standardise the weight to a constant age by a simple multiplicative procedure, where Wc is the estimated weight at a constant age C and W is the weight when the animal's age is D. Age adjustment can also be made by including it as a covariate in the least squares analysis.

Sheep and Goats

The procedures that should be used for dairy sheep and goats are similar to those for dairy cattle. For sheep and goats kept for meat and/or fibre, the procedures are similar to those for beef cattle. The additional factor that should be considered is the maternal effect due to the type of rearing, the classes being single, twin and twin raised as single. Similarly, age of dam effects may be considered by classifying ewes into 2, 3 and 4-6 year old ewes. Since this effect is most marked between 2-year-old and older ewes, separation of dams into these two categories is often adequate, particularly in small flocks.

Breeding Value

An animal's performance adjusted for environmental effects provides the best estimate of its breeding value where other information is not available. For characters of low heritability, the accuracy of this estimate is low. Information on earlier performance of the animal and records of its relatives can, however, increase the accuracy of the estimate. Males should be evaluated with greater accuracy than females because each male transmits its genes to a larger proportion of the progeny than a female.

Dairy cattle

Bulls: Progeny testing procedures for evaluating bulls are based on the contemporary comparison method in which daughters of a bull are compared with daughters of other bulls milking in the same herd at the same time. The difference between the daughter and contemporary performance is weighted by the heritability (which depends on the numbers of daughters and contemporaries) to get an estimate of the sire's breeding value. Modifications of this procedure such as the cumulative difference (CD), best linear unbiased production (BLUP) and modified contemporary comparison (MCC) are now used to eliminate the bias that may arise due to genetic trend.

All these methods take into account genetic differences between herds and the genetic competition faced by individual sire's daughters. Such sophisticated procedures may not be necessary in most parts of the tropics where progeny testing is in its infancy. However, where a testing programme has been going on for some time, adjustments for genetic trend may be necessary.

An assumption implicit in progeny testing of sires using the contemporary comparison method is that the dams of the contemporaries are a random sample of the population. This may not hold in situations where crossbred (exotic x indigenous) populations are used in progeny testing. The crossbred dams may differ in the percentage of exotic blood and therefore in their breeding values. Sires' breeding values estimated as

$$A = 2\,b\,(Y - H_Y) \text{ should be corrected to}$$

$$A = 2\,b\,(Y - H_Y) - 0.5\,h2\,(D - H_D) \text{ where}$$

Y = yield of daughters, H_Y = yield of contemporaries,

D = yield of dams of daughters and

H_D = yield of dams of contemporaries.

Where AI coverage is not extensive and there is concurrent use of AI and natural service (NS), daughters of AI bulls may milk concurrently with those of NS sires. The latter records could also be used in evaluating AI bulls by including in the least squares model the records of daughters of each young AI bull as a separate subclass and grouping records of all NS bulls' daughters into one subclass.

Cows: The breeding values of cows chosen as dams of bulls should be estimated accurately. This is done by the cow index. The procedure adopted is to first express the cow's record as a deviation from the herd mean so as to eliminate the influence of level of production in the herd. This information is then combined with that of its relatives, particularly those of its paternal daughters (sire's breeding value) and dam's performance to increase the accuracy of estimation of the breeding value. The weighting factors to be used to combine this information are given by Johansson and Rendel. Cow indices are also used in culling of cows but the procedures used are less sophisticated than in culling dams of bulls, since the same degree of accuracy is not required. Most often, the cow record expressed as a deviation from the monthly average of records completed during the last 12 months is adequate.

Beef cattle

Cows: Cows are evaluated on the weight of the weaned calf, in the same manner that dairy cattle are evaluated on milk yield. In the cow index system used by the Meat and Livestock Commission in UK, the superiority in terms of the calf weight over that of its contemporaries is calculated for each calf of a cow. The sum of the contemporary comparisons is then adjusted for the number of calves produced by the cow and the number of contemporaries, to give the weighted cow index. This is analogous to the "most probable producing ability" (MPPA) used in USA.

Reproductive potential of the cow and calf survival are of great economic significance in beef cattle production in the tropics. Wilkins therefore suggests an index that includes these traits. The trait measured is the weight of weaned calf produced per month of productive calving interval, the latter being defined as the interval between the birth of the last calf reared to weaning and the birth of the calf under study. Thus, a calving in which the calf dies before weaning is deemed not to have occurred. The cow index is then calculated by expressing the above measure as a percentage of all cows calving during the month.

Bulls and heifers: Bulls and heifers are usually ranked on post-weaning weight, which is taken at about 18 months of age. For large herds, the least squares analysis gives unbiased estimates adjusted for environmental effects. These can be expressed as deviations from the overall adjusted mean for ranking purposes.

Wilkins describes a procedure suitable for small herds where at least 2-3 bulls are reared per month. All bulls between 14 and 18 months of age are weighed monthly and a linear regression of weight on age is calculated. At each weighing, every bull is given a rating which is the percentage deviation of the bull's weight from the predicted mean weight at that age. The mean of the five ratings (obtained from the 14th to the 18th month) is the bull's index. In this system, the population of bulls under test changes monthly, older bulls going out and young ones coming in. Heifers can also be evaluated by the above procedure. However, since heifer rankings need not be as accurate as those of bulls, evaluation may be made on the basis of a single weighing at 24 months.

Sheep and Goats

The procedures used for meat or wool breeds of sheep are similar to those for beef cattle. The most important characteristic in flocks that lamb seasonally the number of lambs born at each lambing. In the New Zealand national recording scheme the deviation of each ewe is calculated each year from the mean of all ewes of her age, and her average deviation calculated by summing this and dividing of the number of matings. To make records of ewes of different ages comparable, its average deviation is multiplied by a factor which takes into account the number of matings (k) on which the average is based, namely k/1 + (k - 1) t, where t is the repeatability of number of lambs born. Then, if F is the ewe's fertility index, $F = \frac{k}{1+(k-1)t}$ (average deviation in lambs born from mean of contemporaries).

The average deviation is now calculated from the deviations in individual years; it can also be calculated as $(P_D - \bar{\bar{P}}_D)$ where D = n/k (n = observed number of lambs born over k joinings) and $\bar{\bar{P}}_D$ = average number born to ewes of ages up to and including the dam's present age.

Evaluation of rams for fleece or body weight can be made on the rams themselves but, for reproductive rate, measurements on female relatives are needed. The dam's fertility index is the most useful since it does not involve lengthening the generation interval. Turner used $\frac{1}{2}h^2F$ to score rams, h^2 being the heritability. As this score can be positive or negative, Turner made adjustments which resulted in a score of 100 for ewes which had produced an average of one lamb per mating, and values above and below this for better or poorer performances.

The evaluation procedures for dairy sheep and goats are basically similar to those for dairy cattle.

6

Selection Programmes in Genetics

In the developed countries, genetic improvement programmes for livestock have advanced to the present stage, chiefly because the infrastructural elements necessary for planned breeding programmes such as intensive performance recording and AI. In addition advances have also been made in population genetics and statistics which have made it possible to develop genetic improvement schemes applicable on a national scale. In contrast, infrastructural facilities for national breeding programmes are lacking or are inadequate in most parts of the tropics.

The need for such facilities has also not arisen because the traditional methods of livestock husbandry practised in these areas, e.g. nomadism or backyard farming, make it impossible to operate a planned breeding scheme as is done in the developed countries. As an alternative, state-sponsored livestock farms have been set up with a view to effecting genetic improvement in these herds or flocks and passing the improved stock, chiefly males, to farmers.

Selection Programmes for Indigenous Breeds

There are many arguments in favour of selection programmes for indigenous breeds of livestock.

i. They are adapted to the tropical environment. Among the more important adaptational qualities are disease and parasite resistance, e.g. to trypanosomes and ticks, the ability to survive and reproduce in hot, humid, environments and the ability to exist on low quality feed and limited water supply. Selection for increased productivity should therefore raise their potential for productive traits without serious detriment to their adaptational qualities.

ii. Germ plasm of exotic breeds is imported into many tropical countries either as live animals or as frozen semen. Crossbreeding of these exotic breeds with indigenous stock poses a danger of extinction of the latter types. In these situations, one can ensure the survival of the local strains only by raising their genetic merit so as to minimise the competition.

iii. Selection among indigenous breeds is an integral part of some crossbreeding programmes such as rotational crossbreeding where exotic and indigenous breeds are used in rotation.

In formulating selection programmes in institutional herds in the tropics, it is important to ensure that they are simple in design and that the traits selected are those of real economic value. Among these, adaptability traits should always be emphasised. It is also important that the population under selection should be exposed to environmental conditions similar to those in which the offspring are expected to perform.

For example, if the objective of selection in an experimental herd of beef cattle is to supply genetically superior bulls to farmers who keep their animals on range without supplementary feeding, it would be inappropriate to keep the experimental animals on high levels of concentrates. This is important to avoid errors in selection caused by genotype-environment interaction. The same argument also applies to experimental herds that observe stringent parasite control measures such as frequent treatment with acaricides or anthelmintics. Such considerations should, however, not lead to the other extreme of imposing excessively harsh conditions that would mask the expression of genetic variation.

Improvement programmes in many institutional farms in the tropics are at present inefficient, as they are confined to small or medium-sized government herds and are often hampered by a lack of continuity in the pursuit of selection objectives. Their effectiveness should be increased by organising integrated programmes that embrace all government farms of the same breed of livestock and also private herds above a certain level of management. Establishing links with private farmers is important not only to extend the genetic improvement to their herds but also to increase the effective size of the population under selection and to ensure that testing of animals is also done under field conditions.

Selection Procedures

The two major selection procedures used in ruminant livestock are performance testing and progeny testing. The former is used for characters of high heritability and easily measured in both sexes, e.g. growth rate. For characters that have a low heritability or are measureable in only one sex (e.g. milk yield) or after slaughter (e.g. carcass quality), progeny tests are employed. Progeny testing, when based on adequate information, gives an accurate estimate of breeding value and is therefore preferred when an animal is to be used widely for breeding as in AI.

Information on other relatives is often used as a selection aid, particularly in choosing young males for progeny testing, e.g. dairy cattle. For this purpose, the information that is most valuable is that of the sire and dam. The additional gain in accuracy by using information from ancestors further back in the pedigree is relatively small. The above principles of genetic improvement form the basis of different types of selection schemes. These may operate within individual herds, among a group of cooperating farmers, or nationally. Some of the schemes are follows.

— *Testing stations*: Animals from different herds are assembled in testing stations and measured under a uniform environment. The influence of the herd environment is therefore minimised and the population under comparison is vastly increased. It is particularly suitable for situations where a large number of small herds are involved each of which does not have the capacity to carry out proper withinherd comparison. Testing stations can be used either for performance testing, e.g. growth rate, or for progeny testing, e.g. milk yield.

— *Progeny testing schemes*: The progeny test depends for its accuracy on having a large progeny group per sire. Where a number of herds are involved in testing, the maximum efficiency is achieved when each sire has equal number of progeny in each of the herds. The scheme usually operates by having a central station from where semen of the different sires on test is distributed to a network of AI centres which carry out the inseminations.

A modification of this scheme called the reference sire scheme enables comparisons to be made also among bulls that are used only within individual herds. The reference sires, which are usually progeny tested animals and available only by AI, are used alongside the individual herd's own bulls in all herds. The breeding values of home-bred bulls are then estimated in relation to the reference sires. Comparisons among bulls in different herds are made through the linkages established by the reference sires. Thus sire replacements can be made from among all herds.

— *Nucleus or group breeding scheme*: This scheme is based on the principle that in each herd there is a small number of genetically very superior animals which if brought together will form a nucleus whose average genetic merit is far greater than that in any of the contributing herds. The important element in this scheme is therefore for a group of farmers to agree to pool their highperforming animals. Once the nucleus herd is assembled, an efficient system of recording and selection is implemented.

The best males are kept for breeding in the nucleus while other selected males are given to the base herds for breeding. By this means, improvements are quickly spread throughout the group. The nucleus may remain open to animals from the base herds, the best females from the latter being admitted periodically and compared with those in the nucleus. Usually, only females are transferred from the base to the nucleus since sire selection will not be practicable in base herds due to managerial reasons.

The main advantage in the nucleus scheme is that the genetic superiority of sire replacements coming into the base herds from the nucleus is far greater than what is achievable in each of the base herds. It is particularly attractive in situations where within-herd selection programmes are ineffective due to small population size or inadequate technical skill.

Since the nucleus breeding scheme shifts the onus of operating the breeding programme from the farmer to the nucleus herd, it seems an attractive method for the tropics because of the limitations discussed earlier. However, the organisation of the scheme may have to be under government control because cooperative ventures among farmers may not always be practicable.

Selection Programmes in Dairy Cattle

Selection Programmes in Tropical Breeds

No serious attempt has hitherto been made with national selection programmes to improve the dairy merit of indigenous cattle. This is partly due to the belief that the limiting factor to increased production is not the genetic merit of the animals but rather the adverse environment, and that improvements in husbandry and veterinary care should precede the setting up of breeding schemes. But when one considers the performance of most indigenous breeds, a remarkable similarity is noticed in their late age at maturity, short lactation, long dry period, low average daily milk yield and lack of dairy temperament. Most of these characters are expressed even under conditions of improved husbandry, thus demonstrating that poor productivity is attributable to low genetic merit. Breeding schemes to raise their genetic potential are therefore fully justified.

The maximum rates of genetic gain in milk yield achievable by selection even in temperate breeds is of the order of 2.0 percent per annum. This low rate of improvement has discouraged implementation of selection schemes for indigenous breeds, since the overall increase in production attainable by selection in cows yielding 500-1,000 kg is not of any great magnitude. It should, however, be recalled that even European breeds were as unproductive as the tropical breeds before the application of selection programmes and it is the application of planned selection programmes that has brought the advance that we see today. Since the techniques for genetic improvement are even more developed today than when they were first used on European cattle, it is not inconceivable that similar achievements could be made in tropical dairy breeds.

Where positive selection efforts have been made to improve tropical breeds, the results have been encouraging. Annual genetic gains of approximately 2.5 percent of the herd mean yield have been reported by Acharya and Lush in Haryana (zebu) cattle in India and by Franklin et al. in the Australian Milking Zebu. The high rate of genetic gain in both populations was partly attributed to the high coefficient of variation for milk yield in the populations, which is characteristic of most dairy cattle in the tropics. It is also clear that, in populations that failed to show significant gains, it was the inefficient selection procedures that were responsible, e.g. East African zebu.

In formulating a breeding programme for dairy cattle in the tropics, the ultimate objectives of the programme and the environment and production system under which these objectives are to be realised should be defined. Thus, the objectives should clarify whether it is milk alone or milk and beef or milk and draught or all three that are to be improved. Further, depending on market demands, the importance to be attached to compositional quality should also be clarified. Similarly, the production system, which can range from nomadism through small sedentary herds to large commercial farms, will determine the importance given to adaptation traits, particularly those concerned with disease resistance and reproduction.

Milk yield is the most important trait considered in selection programmes for dairy cattle, though fat percentage is also measured routinely in several countries. Since the negative genetic correlation between milk yield and fat percentage may result in the latter declining to low levels, it is usually monitored to eliminate undesirable animals. Bull fertility is important, particularly in AI bulls, while in the female reproductive performance is considered in culling, this being done more for economic reasons than for genetic improvement.

The principal selection method employed is progeny testing where the breeding value of bulls is assessed on the performance of their daughters. The test depends for its efficiency on a large progeny group per sire and on having each sire represented in several herds. AI is necessary to separate environmental from genetic effects. Milk yield records are usually restricted to those from first-calving heifers since this gives the largest group of unselected daughters, although some countries, e.g. USA, consider records of older animals as well.

Young bulls that enter the progeny test are bred from the best progeny-tested sires and selected dams. Each young bull is used on about 500 cows in milk-recorded herds to ensure that first lactation records of at least 80 daughters are obtained. Matings are done at random to ensure that the dam contribution to progeny genotype is similar for all sire progeny groups.

Though this type of selection programme appears to require a great deal of infrastructural facilities, an efficient testing programme could operate even with a simpler organisation if recording is limited to that required for progeny testing. Thus, Ireland operates a scheme which tests 40 bulls annually on approximately 75 daughters per bull. Only 3 000 daughter records are necessary, though the scheme serves a population of approximately one million cows.

Institutional herds in the tropics

Institutional herds have an important role in the tropics since these may be the only herds where genetic improvement of indigenous stock can be carried out. Institutional facilities are also necessary in selection programmes for exotic breeds which are required in pure or crossbreeding programmes. Artificial insemination should be employed for breeding even in institutional herds so that large sire progeny groups are obtained for accurate progeny tests.

Where the institutional herd is too small, it is necessary to enlist the support of the progressive farmers in the area to participate in the progeny testing scheme. In situations where only natural mating can be employed, progeny testing loses much of its efficiency since the number of bulls required is so large that the selection intensity as well as the accuracy of the test are much reduced. Selection of bulls on their dams' records will then be equally efficient provided that bulls are replaced annually to minimise the generation interval.

An example of a successful breeding programme involving an institutional herd is that used in the Sahiwal herd in Naivasha, Kenya. The breeding plan given below is based on the description by Meyn and Wilkins whose schematic diagram is also reproduced in Figure 1. The

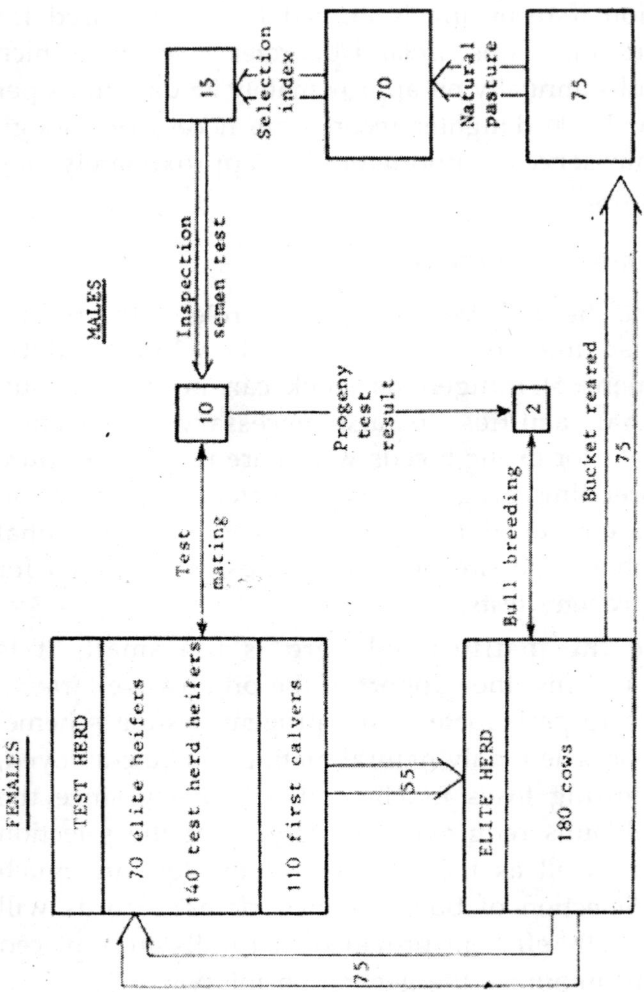

Figure 1. Breeding plan at the National Sahiwal Stud, Naivasha, Kenya.

herd consists of about 500 cows. Breeding is by AI with semen from the best two progeny tested bulls. Female selection is almost entirely on milk yield during the first two lactations. About 50 percent of the heifers that enter the milking herd annually are culled at the end of the first lactation and a further 50 percent after the second lactation. Thus only 25 percent of the heifers freshening annually are kept to the third and higher parities. These animals constitute the elite herd consisting of about 180 animals. They are bred only by progeny tested bulls while the balance of the herd (320 animals) consisting of heifers and first calvers is used for progeny testing young bulls.

Bulls are selected from among those born in the elite herd. They are first performance tested for growth rate on pasture to 2 years of age and 15 bulls (out of the 70 born annually) are chosen on an index that combines growth rate with dam's milk yield. A further five animals are removed for poor physical conformation and semen quality and the remaining 10 are progeny tested. They are bred to females outside the elite group (about 320 animals) and each bull is evaluated on the first lactation yields of about 14 daughters. The best two bulls are chosen for use in the elite herd and national AI scheme. The genetic progress with this type of programme is estimated at 3-4 percent per year.

New breed formation

In programmes involving crossing of exotic and indigenous stocks for developing a new breed, selection has an important role to play after the initial crossings are made to establish the desired gene pool. Selection can be not only for the improvement of production traits but for adaptability traits as well.

Selection for milk yield: The procedures used in selection for milk yield involve progeny testing. Testing of bulls is, however, not possible in the early generations of new breed formation (F1 or backcross depending on the composition of the new breed since young bulls are required to mate with heifers of thier own age and generation. Further, even from a genetic viewpoint, selection should commence only after segregation and recombination of the gene combination's of the parental breeds have occurred. The opportunity for this to happen will arise after inter se mating starts. Progeny testing of bulls should therefore commence with bulls born after this stage.

Private farmer participation is important in sire evaluation for reasons given earlier but complications may arise if the dams mated to the sires differ in their genetic make-up. This is usually the case in situations where crossbreeding programmes in private herds have been under way for a number of years before selection schemes commence. The crossbred dams may then differ in respect of both breed and grade of exotic blood and bias the progeny test due to their different contributions to the progeny. This was the case in Brazil when progeny testing of crossbred bulls was initiated in 1977.

The approach adopted to minimise the bias in this programme was to use all bulls on all farms to avoid confounding of dam breed type and bull. Further, within farm and year, dams were balanced among bulls according to percent European blood and progeny lactations were expressed as deviations from herd-year-season means corrected for age and grade of European blood. This method was suitable, because herd sizes were large (40-700 cows). It is clearly unsuitable where herd sizes are small. In such situations, the concept of the herd as a unit may have to be abandoned and comparisons done within groups of herds, grouped on the basis of average grade and production level.

Adaptability traits: Selection for adaptability traits should wherever possible proceed simultaneously with improvement for milk yield. Measurements for these traits may, however, have to be limited to experimental herds because of the facilities required. This was done in the development of the Australian Milking Zebu (AMZ)—a breed derived from the Indian zebu and Jersey. While progeny tests of bulls were done in both experimental and outside herds, selection for adaptability was confined to the experimental herd at Badgery's Creek in New South Wales. Young bulls were first screened for heat tolerance and tick resistance and only 1 in every 6 or 8 bulls screened was finally progeny tested.

The heat tolerance test was designed to select bulls that maintained their body temperature under heat stress while simultaneously having a high energy conversion rate. Animals were tested in climatic chambers and measurements recorded of physiological changes (rectal temperature, sweating rate and respiration rate) and feed and water intake when they were moved from low to high temperatures (from dry and wet bulb of 21° and 18°C to 38° and 24°C respectively).

Body temperature and respiration rate were given a negative score and sweating rate and feed conversion given a positive score. The best third of the animals chosen on this test were put through tick challenges. This was done by artificially infesting each animal with 40,000 larvae of Boophilus microplus and counting the number surviving to adult stage from the 18th to the 22nd day after infection. The best 50 percent of the animals that had the least number of engorged ticks were finally chosen for progeny testing after examination of semen quality.

Absence of Milk Recording Programmes

In the absence of AI and milk recording programmes, institutional herds offer the only opportunity to carry out recording and selection. Cunningham has proposed a kind of open nucleus breeding scheme (Figure 2) that may be suitable for this situation. The plan operates around a central herd which should preferably be under government control and carry out proper recording and breeding practices. The supporting base population is the village herds which provide cows to replace about 10 percent of the cows in the central herd annually. Selection from village herds will be by simple procedures involving judgement by eye, milking ability, size, conformation and condition. Where feasible, test milkings may also be done. The selection of these animals will be done by officers who would tour many villages to select the best animals from as wide a population as possible.

Bulls are bred from the best cows and selected bulls in the central herd. They are then evaluated on their own growth rates and dams' and milk yields and the best are chosen as sires. The lowest yielding 10 percent of the cows are replaced with new animals drafted from village herds. The latter animals are recorded during the following year with the remainder of the central herd and re-evaluated. The benefits to the cooperating farmers from this scheme will accrue from the sale of selected bulls from the central herd.

Assuming that the central herd has 200 cows and that the best third of the bulls born annually are chosen, about 20-25 bulls will be available for sale to farmers after meeting the central herd's requirements. This may satisfy the requirements of the cooperating farmers but will have hardly any impact among farmers outside the scheme. This is a serious shortcoming of programmes involving

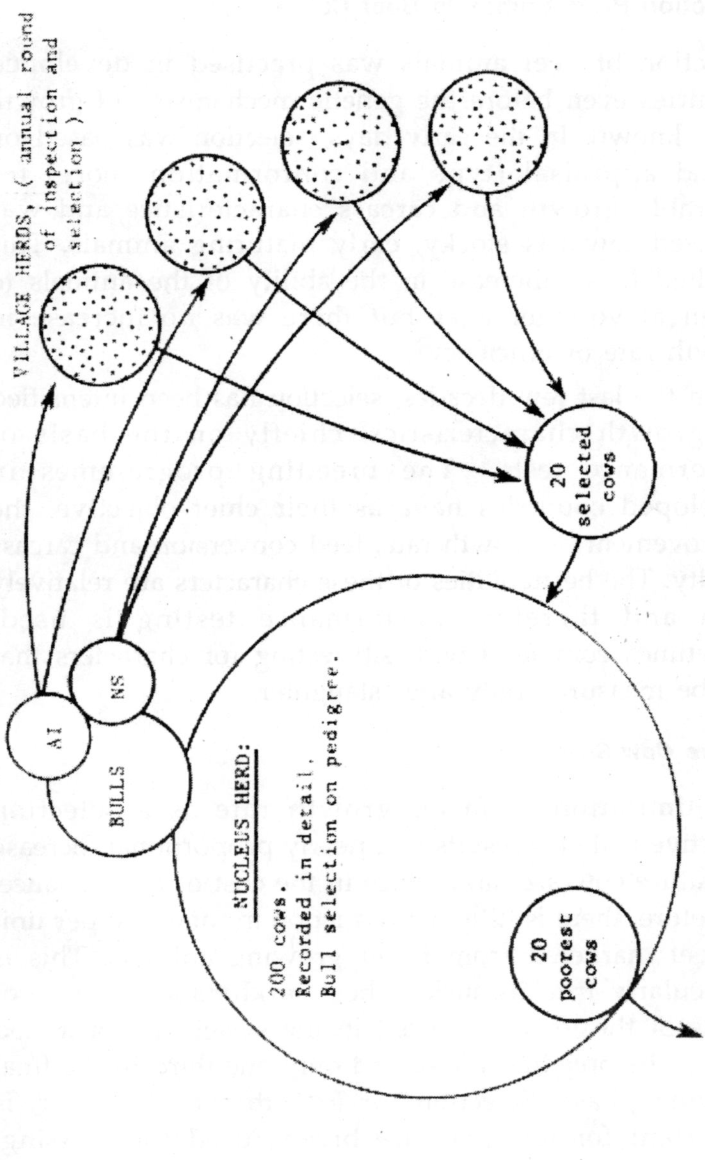

Figure 2. Open nucleus breeding scheme

central herds where supporting AI services are inadequate.

Selection Programmes in Beef Cattle

Selection of beef animals was practised in developed countries even before the genetic mechanisms of growth were known. In the early days, selection was based on visual appraisal (type and conformation score) for desirable growth and carcass characteristics and was directed towards stocky, early maturing animals. This resulted in an increase in the ability of the animals to fatten at younger ages but there was no increase in growth rate or efficiency.

In the last few decades, selection has been intensified for growth characteristics, chiefly on the basis of performance tests. The breeding programmes in developed countries have as their chief objective, the improvement of growth rate, feed conversion and carcass quality. The heritabilities of these characters are relatively high and therefore performance testing is used, sometimes combined with sib testing for characters that can be measured only after slaughter.

Mature Cow Size

The limitation of faster growth rate as a selection objective is that it results in a nearly proportional increase in mature cow size and hence in the cost of maintenance. Therefore, there is little net reduction in total cost per unit of beef marketed from faster growing animals. This is particularly true in suckler beef production, since two-thirds of the total feed used in the system is consumed during the breeding phase and only one-third in the final growing phase. Selection for growth rate, however, is important for terminal sire breeds used for crossing, where the antagonism due to the higher dam size does not arise.

Table 1. Calculation of productivity indices

Parameter	Code	Calculation
Cow mortality during year (%)	A	
Calving percentage (%)	B	
Calf mortality to 1 year (%)	C	
Percent of calves reaching 1 year (%)	D	B (100-C)÷100
Calf weight at 1 year (kg)	E	
Annual milked-out lactation yield (kg)	F	
Percent of cows completing a lactation (%)	G	100 – (C÷2)[1]
Total liveweight equivalent of milked-out yield (kg)	H	F (G÷100)÷9
Total weight of 1-year-old calf produced per cow (kg)	I	E (D÷100)
Weight of 1-year-old calf plus liveweight equivalent of milk produced per cow maintained (kg)	J	(I+H)÷(100÷A÷2) ÷ 100[2]
Average cow body weight (kg)	K	
Weight of 1-year-old calf plus liveweight equivalent of milk produced per 100 kg of cow maintained annually (kg)		J×100÷K

[1] A cow whose calf dies during a lactation period is considered on average to have produced milk during half the period.

[2] Cows dying during the year are considered on average to have been maintained for half a year.

Similar arguments have also been put forward to show that larger breeds of beef cattle are not intrinsically more efficient than small breeds. This fact is relevant to the situation in some tropical countries where cattle breeds differ greatly in size (such as zebu and trypanotolerant breeds in West Africa) and the smaller breeds are often considered inefficient. Such comparisons based on size alone can be very misleading since efficiency of production depends on many other factors.

FAO/ILCA/UNEP have proposed breed comparisons on the basis of a cow productivity index (Table 1) that combines reproductive rate, cow and calf viability, milk yield, calf growth and cow weights. The index is an estimate of the total weight of a one-year-old calf plus the liveweight equivelant of milk produced per unit weight of cow per year. The merit of such an index lies in relating all the important production characters to the weight of breeding cows which determines maintenance costs.

When large zebu breeds and the small trypanotolerant breeds were compared by this index, virtually no difference was found in terms of productivity per 100 kg cow maintained per year, or in feedlot performance estimated as weight gain per 100 kg body weight maintained. Small breeds are therefore equally efficient for beef production as the larger breeds, but the optimum weight of slaughter for maximum efficiency will be lower for smaller breeds. Smaller breeds may also be at an advantage under conditions of extreme fluctuations in feed supply. They seem to have the ability to lay down fat on a lower plane of nutrition which probably explains their ability to remain in a better condition than larger breeds during the dry season when feed is scarce.

Selection Schemes

In spite of the shortcomings of growth rate as a selection criterion, it continues to be used as the major selection objective in many developed countries. Some of the important breeding schemes in these countries are discussed and their relevance to the tropical situation examined later.

a) *Central performance tests*: Male calves with above average weaning weights in their herds are usually brought in groups at 150-190 days of age for performance testing. At the test station, they remain for a period of 7 months till they are 400 days old during which period they are given a test ration and various measurements are recorded. Even though herd influences are supposed to be eliminated in such central tests, the carry-over effects of pre-test management can bias test results due to compensatory growth. Such influences can be minimised by (i) allowing animals that enter the test at 5-6 months to undergo an adjustment period of 3-10 weeks on the test ration before beginning the official test which should last 5-6 months or (ii) starting the performance test at a younger age of about 90-120 days and terminating at about 400 days of age. The type of feed and the method of feeding vary between test stations.

In general, roughage is offered ad libitum, and the amount of concentrates varied according to the weight of the animal so as to keep the roughage: concentrate ratio constant irrespective of growth rate. The test period should be long enough to bring out the genetic variation between animals. This depends on the type of ration. Thus, while 5-6 months is adequate for a high concentrate ration, it should be longer if the ration is high in roughage. For animals that are only on pasture, a period of 12 months is advisable.

b) *Selection within herds*: This scheme operates by identifying cows with superior mothering ability and calves with superior growth rates. Calves are weighed at ages close to 100, 200, 400 and 500 days of age. The 100- and 200-day weights are used as indicators of mothering ability while the 400- and 500-day weights are used in selection of bull and heifer calves.

c) *Nucleus breeding scheme*: The principles of this scheme have been discussed earlier.

d) *Progeny testing*: Progeny testing is used only to a limited extent in beef cattle because, for improvement of growth rate, the performance test is cheaper, quicker and can be applied to many bulls. However, progeny testing of beef bulls may be carried out when they are to be used in AI, for traits such as ease of calving and carcass quality. Bulls to be progeny tested are first selected on the basis of a performance test and then mated to cows in commercial herds. Information is collected on dystocia, calf mortality, liveweight gain, carcass weight and carcass quality. In such tests, where the performance test precedes the progeny test, over 80-90 percent of the genetic gain in growth rate is attributable to the performance test.

e) *Reference sires*: Most beef herds use natural service for management reasons. The reference sire scheme may therefore be more relevant in beef than dairy cattle, since only the small number of cows required for insemination need to be observed for oestrus.

Traits for Tropical Beef Cattle

The environment to which beef cattle are exposed differs in most cases from that of dairy animals. The latter are often shielded from environmental stress by provision of shade, improved pasture, supplementary feed and veterinary care. Beef cattle, on the other hand, are usually kept on land of low potential. Their main function is to convert pastoral vegetation into meat while being exposed to the total impact of heat, parasites, infectious diseases, periodic malnutrition and shortage of water. The most important trait required is therefore the ability to survive and reproduce in this environment.

Thus, the indigenous breeds of beef cattle which possess some of these characteristics will have to be relied on to make the major contribution to beef production. Local breeds should be evaluated on their overall productivity before any hasty decisions are made to substitute them by imported temperate or large zebu breeds. Among the important adaptability traits of tropical breeds are disease resistance qualities. The ability of zebu breeds to resist tick infestations and the tolerance shown by certain African breeds such as the N'Dama to trypanosomiasis are well documented. The relationship between these traits and other performance traits is, however, not known. For example, all trypanotolerant breeds are small compared to their susceptible counterparts which suggests that small size is an integral part of tolerance. If an antagonism exists between large size and trypanotolerance, selection of these breeds for growth rate may result in increased susceptibility to trypanosomiasis.

The production system used in most parts of the tropics is one of pure breeding (suckler beef). In this system, most of the female offspring enter the herd as cow replacements and only surplus males are used for fattening. The calves usually suckle their dams for a long period (6-12 months) and are later fattened on low-quality pasture. The sole output from the cow is thus the weaned calf. The important traits in this system are, therefore, reproduction, milking and mothering ability of the cow and growth potential of the calf. These traits are considered below:

- *Fertility*: The reproductive rate in most tropical breeds of cattle is low. Yet, this trait is very important not only in the economics of beef production but also in genetic improvement because of its influence on selection intensity. Though the heritability of the various measures of reproduction is low, rigid culling to eliminate animals with low fertility has improved fertility in zebu cattle. However, as most of the improvement resulting from such selection may be non-genetic, fertility may decline in subsequent generations unless the practice of culling on reproductive performance is sustained.
- *Mothering ability*: As important as fertility in tropical rangelands is mothering ability which determines survival rate and preweaning growth rate of the calf. Increased milk yield may increase weaning weight but it may be detrimental to the overall economics of beef productivity if the subsequent conception of the cow is delayed. To avoid such antagonisms, a cow index that combines reproductive rate, calf survival and weaning weight may be preferred for selection purposes. The heritability of this index, however, has not been estimated.
- *Growth rate*: Selection for growth-rate is usually done on the basis of the final weight at a constant age such as 18 or 24 months. This is preferable to selection on the basis of gain from weaning to a constant age because of inaccuracies that may arise from compensatory growth as discussed earlier. The problem of concurrent increase in mature size

consequent to selection for growth rate is equally applicable to tropical cattle. One of the approaches that has been proposed for temperate breeds to minimise the increase in mature size is selection on the basis of relative growth rate (RGR). It is measured as the daily gain relative to the average size over the time interval tested

$$RGR = \frac{Ln\ W_{t2} - Ln\ W_{t1}}{t_2 - t_1}$$

Where Ln denotes the natural logarithm of weight W at time 1 (t_1) or time 2 (t_2). The effectiveness of this parameter in changing the growth pattern has still to be determined.

— *Longevity*: The length of productive life is important since the relative cost of raising a calf decreases with the increase in its productive life. However, as selection for longevity increases generation interval, males may be selected at younger ages on their own structural soundness and the longevity of close relatives.

Selection Programmes in the Tropics

A clear definition of the environment in which offspring of the selected populations are expected to perform is necessary. This is important because of the wide variation in nutritional level and environment in beef cattle production systems which can range from grazing on natural grasslands to feedlot fattening. Recent Australian work has shown the importance of genotype-environment interactions in growth rate. On a low plane of nutrition, the response to selection is chiefly through a reduction in maintenance requirement which results in low appetite and reduction in inherent growth potential. Conversely, on a high plane of nutrition, selection increases growth rate principally through increased appetite.

Thus, animals selected under either of the extreme conditions of nutritional deprivation or high concentrate feeding will be unsuitable if conditions are reversed. A practical approach when the environment cannot be defined, would be to select animals on a moderate plane of nutrition. This may be grazing on natural or improved pasture during the rainy season and minimum supplementation with conserved forages during the dry season. Genetic improvement programmes similar to those used in developed countries are suitable in the tropics if the herds are large and facilities for performance recording are available. However, in most cases, either herd sizes are too small or where they are large, recording may be impossible, e.g. nomadic herds. Some alternative schemes are discussed below.

Large Organised Ranches or Institutional Herds

Many tropical countries have government-owned herds with the objective of producing superior stock for distribution. Some of the important ingredients for the success of a breeding programme in such herds are:

i. A reasonable population size to permit adequate selection intensity without appreciable increase in inbreeding. A suitable size for a breeding unit that uses about 8-10 bulls per annum selected at an intensity of 1 in 10 is about 200-300 cows.

ii. Short generation intervals: The use of old bulls simply because they are impressive and heavier compared to younger bulls is genetically unsound. If a breeding programme operates efficiently, a bull of superior genetic merit should sire better performing sons and therefore use of a bull for more than one year is likely to reduce the rate of improvement in the long run. Bulls should therefore be replaced rapidly to minimise generation interval. Selected young bulls should be used for breeding as early as possible.

iii. Above average genetic merit of the population. This is important if a closed herd selection scheme is adopted. If the herd is to be opened, selected outside bulls may be introduced to a small part of the herd and their progeny compared with those sired by home-bred bulls.

A selection scheme for beef cattle can operate without the paternity of the calves being known. Where sire information is necessary (to increase the accuracy of estimation of breeding values), well fenced paddocks are required to facilitate single-sire matings.

A limited breeding season of about 8 weeks is an advantage particularly in mediumsize herds (200-300 cows), since calves are raised under uniform conditions and adjustments for environmental effects may be unnecessary. Where calvings occur all the year round, comparisons should be made only among contemporaries if reasonably large groups are available or data are adjusted for environmental effects.

Calves should be weaned at about 6 months and their weight and date of weaning recorded. Cows may be culled at this time on reproductive criteria, soundness, age and calf performance. The weaned calves should be placed on a nutritional regime which, while being adequate for expression of genetic variation, is not too different from commercial practice. They should be weighed at about 18 months of age and ranked on the basis of the adjusted weights. Tests for heat tolerance and tick resistance may be incorporated at this stage using similar procedures as in dairy cattle. Selection in females is made on the basis of 18-month weight but animals that fail to conceive during the breeding season (if a limited season is used), or within a specified period after they reach a suitable breeding weight, should also be culled. Matings among sibs, and if possible, among more distant relatives should be avoided.

Over 80 percent of the genetic progress in this type of selection scheme is due to selection of bulls. This can be achieved without much record keeping. The only data required are date of birth, dates of weighing and weaning and 18-month weights. No records are accumulated on cows, these being culled on their current performance only. Ignoring age-of-dam effects may not cause any serious bias in the ranking of bulls on 18-month weight except perhaps in bulls produced by primiparous dams.

Nucleus Breeding Scheme

This scheme seems attractive for the tropics because of the limited scope for recording. In most instances, the only data available may be those obtained from questioning of farmers. Thus animals selected from the base herds may be chosen on the maternal criteria of regularity of calving and calf survival. An "open nucleus" approach may be used in selection for disease resistance such as trypanotolerance since the degree of exposure to disease is greater in base herds. Animals chosen from them may therefore be genetically superior to those in the nucleus, if the degree of disease challenge in the latter is low.

Central Testing Stations

A test station in the tropics to measure growth rate can

consist simply of a grazing area with the necessary weighing scales, a cattle crush and other ancillary equipment. In many tropical countries, these can be established without much capital outlay by converting those government breeding stations that are too small in terms of herd size to be useful for selection programmes. Test stations should preferably be set up in areas representative of the environment in which animals are kept. This is important in breeds selected for disease resistance so that productivity can be measured under conditions of natural disease challenge. The problem of inbreeding is severe in some tropical herds because not only is one bull used at a time but entry of outside bulls is also not permitted. Farmers should therefore be encouraged to exchange their tested bulls.

Dual Testing

Dairy breeds are sometimes selected for both milk and beef as in many European countries where dairy animals are an important source of beef. In testing for both traits, the dairy progeny test is preceded by the performance test and between a third and a half of the young bulls are culled. The remaining bulls are then progeny tested for milk yield by the usual procedure. There is no major genetic antagonism between milk yield and growth rate but selection for increased growth rate would increase mature size. The advantage in terms of growth rate has therefore to be offset against the increased maintenance requirement of the larger dairy animal.

Selection Programmes in Sheep and Goats

All breeds of sheep are likely to produce surplus animals for meat, but with some the main emphasis is on wool, with some on milk, while others again are dual- or even triple-purpose. Genetic improvement work in sheep in developed countries in the early days concentrated on growth and meat conformation, or on wool weight and quality. It has been realised, however, that reproductive rate is the most important factor, whether wool or meat is the main product and that selection on high wool or body weight without adjustment for litter size at birth could lead to selection against twins and so a lowering of reproductive rate. This realisation has led to selection programmes aimed at increasing the reproductive rate of pure breeds and to crossing programmes with prolific breeds such as the Finn.

Sheep Improvement Programmes

The principles and the methods used are similar to those for beef cattle.

a. *Within-flock selection*: One of the well operated schemes for selection within flocks is the Sheeplan in New Zealand. It involves a recording scheme in which measurements are taken of four important traits of productivity—number of lambs born, weaning weight, weight at 15-16 months of age and fleece weight at the same age. The first trait is mandatory for all users while the others are optional. The data are adjusted for environmental effects and selection lists made up for each flock. The criteria used are weaning and 15-16 month weights for meat breeds used in terminal crossing while for dual-purpose breeds all four traits are used. They may be combined into an aggregate index by methods discussed earlier.

b. *Central testing stations*: Performance testing stations are not used to the same extent in sheep as in beef cattle. Where they are used, ram lambs are tested and ranked for final body weight and fleece weight.

Some testing stations are also used for progeny testing, e.g. in the Federal Republic of Germany. Progeny groups are brought in as lambs at 60 days of age and assessed for weight gain, feed conversion and carcass quality.

c. *Progeny testing*: Improvement of milk yield in dairy sheep and goats requires progeny testing. It is used in some countries, e.g. France and Israel, but it is only where AI is widespread that sufficient progeny records can be obtained for a satisfactory progeny test. Progeny testing by a novel procedure described as the "ram circle" and specially designed for small flocks is used in Norway for improvement of growth rate, carcass quality and wool. Several neighbouring farms cooperate to form a ram circle of about 300-400 ewes in which about 10 rams are progeny tested. Each flock gets one ram on the first mating day and starting from the second day, the rams are circulated each day till the end of the mating season. By this means a fairly accurate estimate of the breeding values of rams bred in different flocks is obtained.

d. *Nucleus breeding schemes*: These schemes, also known as cooperative schemes, are becoming more widespread in New Zealand and are also being tried out in Australia. Some comparative studies of the genetic progress between nucleus programmes and conventional selection schemes made in New Zealand seem to bear out the superiority of the former, as expected on theoretical grounds.

Selection of Sheep and Goats

Tropical sheep breeds grow a coat of hair with little or no wool, or they produce coarse, carpet-wool fleeces containing medullated fibres and kemp. They are exploited for wool, meat, skins, pelts and sometimes milk. The market for fat lamb is poor, but there is a growing demand for mutton from older animals. Lean meat is often preferred though in some countries the fat tail of indigenous sheep is still considered a delicacy. The coarse fleece is used for making carpets, rugs, blankets, tents or even local clothing. There is a high demand for carpet wool in world markets which warrants the keeping of these tropical carpet breeds and improving their productivity. Skins from tropical sheep are used for manufactured goods (e.g. gloves), while the pelt from one particular breed, the Karakul, is highly prized as Persian lamb.

An important characteristic of tropical sheep is their adaptation to the environment. The harsh, arid tropical environment exerts a profound influence on the productivity of sheep. Nutritional deficiencies and climatic extremes impose a restriction on production during much of the year. Thus, the most important economic characteristic required in tropical sheep kept under current management systems is adaptability to the environment. Tropical breeds have acquired this characteristic through a long period of natural selection and therefore sheep improvement programmes in the tropics should rely largely on genetic improvement of adapted breeds.

In formulating breeding schemes, certain flock characteristics inherent in the production systems in the tropics should be recognised. In general, flock sizes are large in breeds kept extensively, but recording in these flocks is often impractical. On the other hand, the flock sizes of breeds kept semi-intensively in backyards are too small for effective selection procedures. Some of the most prolific breeds of sheep are kept under these systems, e.g.

D'man of Morocco, Priangan of Java, Barbados Blackbelly of tropical America. Breeding programmes should therefore give consideration to selection in institutional flocks or group breeding schemes.

Selection for Reproductive Rate

Among the selection objectives for sheep, reproductive rate has the highest priority, as previously discussed, whether animals are raised for meat, wool or skins. Reproductive rate may not be as important where milk is the main product and lambs are removed early, except that a higher reproductive rate contributes to increased selection intensity. Reproductive rate can be increased either by raising the number of offspring at a parturition, or by decreasing the period between parturitions. In tropical countries, where breeding seasons are long, and nutritional levels frequently low, the second may be the more practical course, as a dam might find difficulty in rearing more than one lamb.

Where multiple births are considered, selection for these is more likely to increase reproductive rate than selection against barrenness. Traits associated with reproduction are usually considered to be lowly heritable, but in spite of this, response to selection has been obtained. Failure to lamb at the first lambing in some sheep breeds may not indicate poor performance; selection should be based on more than one record. On the other hand, ewes which produce more than one lamb at their first lambing are likely to be good performers.

The simplest method of selection for reproductive rate is to select rams from among the multiple-born lambs on the basis of other criteria such as growth rate or wool weight. In the case of ewe lambs, all lambs born in multiple births are chosen and the balance required is made up from single lambs. Ear notching or tattooing may be done to identify multiple-born lambs if they are not individually identified. When full lambing and pedigree records are available, lambs are selected on the basis of a dam index which combines data of all her lambing records.

The genetic correlation between prolificacy and growth rate is positive and therefore selection for both traits can be done simultaneously without any danger of antagonism. Where selection is for weaning weight, the latter should be adjusted for the effects of age at weaning, type of birth and age of dam. However, when the trait selected is weight at a later age, such as at 15-16 months, adjustments are unnecessary as these weights are not much influenced by the above factors.

Prolificacy and growth rate

The breeding scheme to be described is suitable for institutional and other flocks that have at least 100 breeding females. Even in flocks of this size, it is necessary to have a short lambing season to avoid the need for adjustment factors which cannot be properly estimated if lambing is year round. A minimum of eight rams should be used if the flock is to remain closed. To minimise the generation interval, rams should be chosen as soon as they reach puberty and replaced every mating season. Assuming that the number of breeding females is about 100 and a percentage of lambs weaned to ewes joined of 100 (a figure likely to be achieved with about 20 percent twinning), the number of lambs of each sex available at weaning is about 50. The first selection of lambs should be on weaning weight.

Though selection of all ram lambs from twins is desirable, a few singles may also be chosen to put some selection pressure on growth rate. Thus, at weaning, all twin and about 4-6 single ram lambs are retained and at 15-16 months of age, when the final selection of 8 rams is made, 6 could be twins and 2 singles. Selection of ewe lambs is less intense because of the need to maintain flock size. Thus, all twin-born ewe lambs are retained and the balance made up from singles based on 15-16 month weight.

Nucleus breeding scheme

The nucleus flock should aim to have about 500 ewes. Such a size is necessary not only to make sizeable selection gains but also to meet the requirements of farmers outside the nucleus scheme. Selection of animals for the nucleus should be based on observation and interviews in the absence of objective data. Thus, ewes selected should at the time of selection be rearing twin lambs and have a history of twinning in one or more earlier lambings. Ram lambs should be selected from among multiple-born lambs on size and constitution. Where the choice of base flocks is wide, it is preferable to select rams from flocks that do not contribute ewes, in order to minimise inbreeding.

Selection for milk and wool

The selection schemes outlined so far have been confined largely to meat production. However, prolificacy is as important in wool as in meat production. The selection procedures in wool breeds would be essentially similar except that fleece weight and quality at 15-16 months of age should be substituted for weight as a selection criterion. The procedures of selection for dairy sheep and goats are similar to those for dairy cattle.

Selection in goats

Goats have received less attention in genetic improvement programmes even though they are more numerous than sheep in some countries. Some of the reasons lie in the difficulties associated with raising goats in large numbers in limited grazing areas as in institutional farms. When kept in this manner, they are more susceptible than sheep to parasites and infectious diseases. The selection schemes proposed for improvement of prolificacy and growth or milk in sheep are also applicable to goats.

7

Genetically Modified Food Crops

Genetic engineering or recombinant DNA technology has been found by plant breeders to be a very precise tool in crop improvement. It has helped in breeding for higher yields, resistance to pests and diseases, tolerance to both biotic and abiotic stresses, improved nutritional quality, and improved processing, storage and post-harvest qualities.

The adoption of genetically modified (GM) crops has been phenomenal from 1.7 million hectares in 1996 to 52.6 million hectares in 2001. This represents an increase of more than 30-fold in only six years. Globally, the principal GM crops are soybean, maize, cotton and canola. Other GM crops grown commercially are potato, papaya and squash.

However, in spite of perceived benefits from GM crops, several health and environmental concerns have been raised. As a result, there is widespread opposition among consumers, and a general lack of public acceptance. On the other hand, pharmaceutical products produced by the same procedures have generally been accepted.

Regulation of GM Crops

Field Testing and Commercialization

Countries which are undertaking research and development which use modern tools of biotechnology have established mechanisms to regulate such activities. For example, in the Philippines, an inter-agency committee created by a Presidential Executive Order, the National Committee on Biosafety of the Philippines, deals with biosafety issues, including field testing. Commercialization of GM crops, however, as well as their importation, are under the supervision of the Department of Agriculture. Biosafety guidelines pertaining to research and development, importation and commercia-lization of GM crops are now in place.

Safety Assessment Procedures for GM Crops

Safety assessment procedures for GM food and food products have been studied carefully since the early 1990's. The United Nations Food and Agriculture Organization (FAO), the World Health Organization (WHO) and the organization for Economic Cooperation and Development (OECD) have held consultations on the issue. They have recommended science-based safety assessment strategies for food and food additives produced by genetic engineering. Such strategies have been adopted by various countries, including Australia, Canada, EC countries, Japan, New Zealand, Philippines, and the USA.

In general, the safety assessments are based on substantial equivalence. This means that if a new food or food component is substantially the same as an existing food or food component, it can be treated in the same manner with respect to safety. It should be noted, however, that substantial equivalence is only the starting point in the biosafety assessment procedure. In other words, it helps to identify potential safety and nutritional issues, but it is not the assessment itself. The procedure requires the submission by the applicant of the following information.

— The characteristics of the traditional or parental product or organism, compared to those of the new product;
— The techniques and components used to produce the new product;
— Whether the new product might cause allergic reactions;
— Whether genes might be transferred from the GM plant to microorganisms in the gut of human beings or livestock; and
— Possible toxic effects.

These data are evaluated by panels of scientists from different disciplines. It should be noted, therefore, that safety assessments starts, not with the GM crop or product, but at the research phase: from the choice of genes/proteins (their function, or mechanism of action, should be known), to the source of the gene (ecological or environmental considerations) to be used. The GM crop undergoes a rigid evaluation of its agronomic performance and efficacy. It is subjected to food/food safety assessment tests, as well as environmental safety tests. Furthermore, assessments must be done on a case by case basis.

Evaluation of Food Safety

All the GM crops that have been released onto the market in various countries have undergone rigorous evaluation for food safety. For instance, the European Union requires the following information on GM crops/food for evaluation:

— The specifications of the novel food;

— The effects of production process applied to the novel food;

— A history of the organisms used as the source of the novel food;

— The effects of genetic modification on the properties of the host organism;

— The genetic stability of the genetically modified crop;

— Specificity of the expression of novel genotypes;

— The ability of genes from the GM crop to survive in, and colonize, the human gut;

— The anticipated intake or use of the novel food;

— Information about any effects of previous human exposure to the novel food or its source;

— Nutritional information about the novel food; and

— Information about the possible toxic effects of the novel food.

Many countries have approved different GM crops. For example, Canada issued letters for 42 GM products, indicating that it had no health concerns over these products. Japan has approved about 40 different GM products although no GM foods are yet on sale in Japan. The EU has approved various GM crops, such as soybean, maize, cotton, potato and rapeseed, although it has recently passed strict labelling laws governing GM crops used as food and feed.

Food Safety Issues

Food safety issues are based on the possible effects of the genes from two or more species, the "transgenes". There is particular concern where such genes might cause allergies or have toxic effects. There is also concern over the possible effects of the antibiotic selection markers, whether these might induce resistance to antibiotics used as medicine.

There are risks inherent to the technology but these are amenable to science-based evaluation. It is, however, a fact that these traits are not unique to GM crops. Many foods contain allergens and toxins. For example, peanuts, seafoods and even wheat and rice contain allergens which affect many people. Cassava, almonds, and sorghum contain cyanogenic glycosides which produce cyanide, and cause death or nerve damage. Legumes contain trypsin inhibitors, lectins and saponins, all of which may have a toxic effect. Many people are intolerant to milk. All these risks have been managed through the years. However, the only treatment for food allergies is dietary avoidance. This depends on being able to identify the source of exposure.

A study by a panel of experts convened by the International Life Sciences Institute (ILSI) and the International Food Biotechnology Council developed a decision tree for the assessment of allergenic potentials of foods derived by modern biotechnology. Some of the general properties of proteins which do not have cause allergies are as follows:

— No history of allergenicity;

— Different in structure or character from known allergens;

— Rapidly digestible;

— Unstable when heated (cooked); and

— Expressed at low levels, compared to the expression of major allergens.

The risks posed by the allergenic potentials of GM foods should be weighed against their potential benefits. As in many conventionally bred crops that contain toxicants and allergens, their risks should be managed. All GM crops which have been commercially released have undergone and passed rigid food safety tests. For example, the insect-resistant and the herbicide-resistant soybeans and maize have been found to cause no oral toxicity in mice.

Many animal feeding studies, using chickens, cattle, sheep and growing bull calves, showed that animals fed with GM crops had the same performance as those fed with the conventional counterparts. Furthermore, selection markers such as the kanamycin resistance gene and its product are digestible and are not allergenic. Another type of risk related to GM crops is that which is a result of the political and social context in which a technology is used. This includes unequal access to the new technology.

The economic gap between the technology users and the non-users, and further disparity in income between rich and poor farmers. This risk can be addressed by developing technologies which are expressly designed to meet the needs of the poor and instituting measures that will provide them affordable access to the new technology. Leisinger contends that this kind of risk occurs because of the existence of a gap between human scientific technical ability, and human willingness to shoulder moral and political responsibilities.

Labelling of GM Crops

Another important concern is the segregation and labelling of GM crops and GM-derived foods. There is a world-wide debate over whether we need to legally require the labelling of GM crops, and food or feed derived from GM crops. The prevailing position in Canada and United States is that if the GM crop is substantially similar to its conventional counterpart, there is no need to label it. However, there is a powerful lobby to require that all GM crop be labeled, so that consumers can exercise their right to choose. The EU has recently passed legislation requiring that all food which contains more than 1% GM materials must be labelled. Japan has similar legislation, but with a threshold of 5%.

8

Transgenic Crops

The overall economic impacts of transgenic crops will depend on a wide range of factors including, among others, the impact of the technology on agronomic practices and yields, consumers' willingness to buy foods and other products derived from transgenic crops, and regulatory requirements and associated costs. In the longer term, other factors such as industry concentration in the production and marketing of transgenic crop technology may also influence the level and distribution of economic benefits.

Farmers who adopt the new technology, especially those who adopt early, may reap benefits in terms of lower production costs and/or higher output. Other farmers could be placed at a competitive disadvantage depending on how consumer preferences and regulatory regimes evolve. If consumers are generally accepting transgenic crops and regulatory requirements are not too onerous, adopting farmers would gain and non-adopting farmers would lose. If consumer opposition grows, however, non-adopting farmers could turn that into a competitive advantage and command a price premium for non-GM products.

Consumers generally benefit from technological innovation in agriculture as a result of lower prices and/or higher quality of the products they buy. The case is more complicated with transgenic crops for at least two reasons. First, regulatory requirements such as mandatory labelling and market segregation could add to the costs of producing and marketing transgenic crops and prevent consumer prices from falling. On the other hand, some consumers are strongly opposed to the technology. These consumers could experience a welfare loss if they were forced either to consume products derived from transgenics or to buy higher priced organic products in order to avoid them.

The net economic impact of transgenic crops on society is thus a highly complex and dynamic concept that is not easily measured. In the first instance, transgenic crops will only be widely adopted if they provide economic benefits for farmers. For developing countries, in particular, a number of economic and institutional factors affect the farm-level profitability of transgenic crops in addition to their purely agronomic characteristics.

Economic research is beginning to show that transgenic crops can generate farm-level benefits where they address serious production problems and where farmers have access to the new technologies. Thus far, however, these conditions are only being met in a handful of countries. These countries have been able to make use of the private-sector innovations developed for temperate crops in the North. Furthermore, these countries all have relatively well-developed national agricultural research systems, biosafety regulatory procedures, intellectual property rights regimes and local input markets. Countries lacking these prerequisites may be excluded from the gene revolution.

The impacts of transgenic crops in developing countries is quite limited, primarily because these crops have been grown for only a few years and in a few countries. This occurs because early adopters achieve a cost advantage over other farmers, earning a premium for their innovation. As more farmers adopt the technology, the cost reduction eventually translates into a price decline for the product that means, while consumers continue to benefit, the gains to farmers decline. A third danger with transgenic crops is that they are, for the most part, controlled by a few large companies. Although these companies do not appear to be extracting monopoly profits from the sales of their products, in the absence of competition and effective regulation, there is no guarantee that they will not do so in the future.

Transgenic cotton is now being grown in a sufficiently large number of countries, under different institutional and market conditions and by different types of farmer, to allow some tentative conclusions to be drawn about the potential benefits and challenges arising from the use of transgenic crops in developing countries. Although it is risky to extrapolate results from one country or one crop to another, the early evidence for transgenic cotton suggests that resource-poor smallholders in developing countries can gain significant benefits from the adoption of transgenic crops in terms of higher and more stable effective yields, lower pesticide costs and reduced health risks from chemical pesticide exposure.

Insect-resistant Cotton

Transgenic cotton containing a gene from the bacterium Bacillus thuringiensis (Bt) that is resistant to certain insect pests was first grown in Australia, Mexico and the United States in 1996 and has subsequently been introduced commercially in six other countries: Argentina, China,

Colombia, India, Indonesia and South Africa (Table 1). Global area planted in Bt- and stacked Bt- and herbicide-tolerant (Bt/HT) cotton varieties increased from less than 1 million ha in 1996 to 4.6 million ha in 2002 (an additional 2.2 million ha of herbicide-tolerant cotton were grown in 2002). Bt and stacked Bt/HT cotton varieties accounted for about 15 percent of global cotton area in 2002 compared with only 2 percent in 1996.

The adoption of Bt cotton has varied greatly across growing regions within China, Mexico, the United States and elsewhere depending on the particular combination of pest control problems. Bt cotton varieties have been rapidly accepted by farmers in areas where bollworms are the primary pest problem, particularly when resistance to chemical pesticides is high. When other pest populations are high, farmers use a mixture of broad-spectrum chemicals that achieve coincidental control of bollworms, reducing the value of Bt control.

Bt cotton

Genes from the common soil bacterium Bacillus thuringiensis (Bt) have been inserted into cotton plants, causing them to produce a protein that is toxic to certain insects. Bt cotton is highly effective in controlling caterpillar pests such as pink bollworm (Pectinophora gossypiella) and cotton bollworm (Helicoverpa zea), and is partially effective in controlling tobacco budworm (Heliothis virescens) and fall armyworm (Spodoptera frugiperda). These pests constitute a major pest control problem in many cotton-growing areas, but other cotton pests such at boll weevil are not susceptible to Bt and continue to require the use of chemical pesticides. As a result, the effect of the introduction of Bt cotton on pesticide usage varies from region to region, depending on the local pest populations.

The first Bt cotton varieties were introduced commercially through a licensing agreement between the gene discoverer, Monsanto, and the leading American cotton germplasm firm, Delta and Pine Land Company (D&PL). These varieties contain the Cry1Ac gene and are commercialised under the trade name Bollgard®. Varieties with transgenes for insect resistance and herbicide tolerance (Bt/HT) stacked together were introduced in the United States in 1997. Monsanto recently received regulatory approval in some markets for a new product that incorporates two Bt genes, *Cry1Ac* and Cry2Ab2. This product, known as Bollgard II®, was commercialised in 2003. The incorporation of two Bt genes is expected to improve the effectiveness of the product and delay the development of resistant pests.

More than 35 different Bt and Bt/HT cotton varieties are on the market in the United States (data from the United States Department of Agriculture [USDA]). These varieties and most Bt varieties worldwide contain genes licensed from Monsanto. An exception is in China, where an independent source of Bt protection is available. The Chinese Academy of Agricultural Sciences (CAAS) developed a modified Bt gene that is a fusion of the Cry1Ac and Cry1Ab genes. In addition, CAAS isolated a gene from cowpea, CpTi, that provides insect resistance through a different mechanism.

CAAS has stacked the CpTi gene with the Bt fusion gene and incorporated them in more than 22 locally adapted varieties for distribution in each of the Chinese provinces. The stacked CAAS varieties are expected to delay the development of resistant pests. The Monsanto Cry1Ac gene is also available in China through at least five varieties developed by D&PL. In Argentina, Mexico, South Africa and elsewhere, the Bt cotton varieties all contain the Monsanto Cry1Ac gene, often in varieties originally developed for the United States market.

Conventional cotton production relies heavily on chemical pesticides to control caterpillars and other insect pests. It is estimated that cotton production consumes about 25 percent of the agricultural pesticides used worldwide, including some of the most toxic chemicals available. Chlorinated hydrocarbons (such as DDT) were widely used in cotton production until these were banned in the 1970s and 1980s for health and environmental reasons. Cotton farmers then replaced DDT with organophosphates, many of which are also highly toxic.

Pests in many regions quickly developed resistance to organophosphates, and pyrethroids, which are less toxic than organophosphates, came into widespread use in the 1980s and 1990s. Resistance to pyrethroids soon developed and multiple chemical resistance has become a severe problem in many growing regions. In areas where bollworms are the major pest and chemical resistance is a problem, Bt cotton varieties have contributed to a dramatic reduction in pesticide use. An important advantage of Bt over chemical control of pests, from a production point of view, is that Bt control is always present in the plant. Because farmers apply chemical controls only after noticing the presence of pests on the cotton plants, some damage will have already occurred.

The effectiveness of chemical insecticide applications, unlike transgenic Bt, also depends on the weather, because rain can wash the chemical away. Bt cotton offers farmers increased certainty of control because it is effective against insects that have developed resistance to available chemical pesticides. As a result, Bt varieties have superior yield performance over a wide range of growing conditions. The estimated difference in yield performance between Bt and conventional cotton varies considerably across time and space because insect infestations vary widely. The relative performance of Bt cotton is highest under conditions where pest pressure is heaviest and chemical pesticide resistance is common.

The major concern associated with the use of Bt cotton is the possibility that pests may develop resistance to Bt as they have with chemical pesticides. This would be a serious problem for organic cotton producers who rely on Bt sprays for pest control. Widespread resistance to Bt would reduce the effectiveness of this option. Pest resistance management is an important part of the regulatory approval process for transgenic cotton.

Transgenic Cotton

The main farm-level economic impacts of the transgenic crops currently being grown are the result of changes in input use and pest damage. Where the new seeds reduce the need for chemical sprays, as can be the case with pesticide-resistant or HT crops, farmers may spend less money on chemicals and less time and effort applying them. Where the new seeds provide more effective protection from weed and pest damage, crops may have higher effective yields. These cost savings and output gains can translate into higher net returns at the farm level. Farm-level economic gains depend on the costs and returns of the new technology compared with those of alternative practices.

The economy-wide and distributional impacts of the introduction of transgenic varieties must also take into account the fact that farmers may expand production as the new technology reduces its costs. This supply response can push prices down, benefiting consumers who may then demand more of the product. As farmers' purchases of seeds and other inputs change, prices for those items may also change, particularly if the input supplier holds a monopoly position in the market. These economy-wide forces will affect the overall level of economic benefits and the distribution of benefits among farmers, consumers and industry.

Table 1. Bt and Bt/HT cotton area, 2001

Country	(000 ha)Area
United States	2 400
China	1 500
Australia	165
Mexico	28
Argentina	9
Indonesia	4
South Africa	30
Total	4,300

In the first year of commercial availability in the United States, Bt cotton was planted on about 850 000 ha or 15 percent of the country's total cotton area. By 2001, 42 percent of the cotton area was planted to Bt and stacked Bt/HT cotton varieties (USDA-AMS, various years). The United States remains the largest producer of Bt and Bt/HT cotton, but its share of global transgenic cotton area fell from about 95 percent in 1996 to about 55 percent in 2001 as adoption in other countries increased.

United States farmers adopted Bt cotton very quickly, especially in the southern states where pest pressure is high and chemical pesticide resistance is most pronounced (Table 2). Bt cotton adoption has had a large impact on pesticide use in the United States. The average number of pesticide applications used against bollworms has fallen from 4.6 in 1992-95 to 0.8 applications in 1999-2001. Carpenter and Gianessi and Gianessi et al. estimate that the average annual use of pesticides on cotton in the United States has been reduced by approximately 1,000 tonnes of active ingredient.

Falck-Zepeda, Traxler and Nelson calculated the annual impacts of Bt cotton adoption in the United States on United States cotton farmers, consumers, germplasm suppliers and foreign farmers for the 1996-98 period using a standard economic surplus model. The estimated amount and distribution of benefits from the introduction of Bt cotton fluctuates from year to year; thus the average figures for the period 1996-98 are also shown in Figure 1.

Table 2. Adoption of Bt cotton by farmers in the United States by state, 1998-2001
(Percentage)

	1998	1999	2000	2001
Alabama	61	76	65	63
Arizona	57	57	56	60
Arkansas	14	21	60	60
California	5	9	6	6
Florida	80	73	75	72
Georgia	47	56	47	43
Louisiana	71	67	81	84
Mississippi	60	66	75	80
Missouri	0	2	5	22
New Mexico	38	32	39	32
North Carolina	4	45	41	52
Oklahoma	2	51	54	58
South Carolina	17	85	70	79
Tennessee	7	60	76	85
Texas	7	13	10	13
Virginia	1	17	41	30

United States cotton farmers gained a total of about US$105 million per year in higher net incomes as a result of Bt adoption, which lowered their production costs and

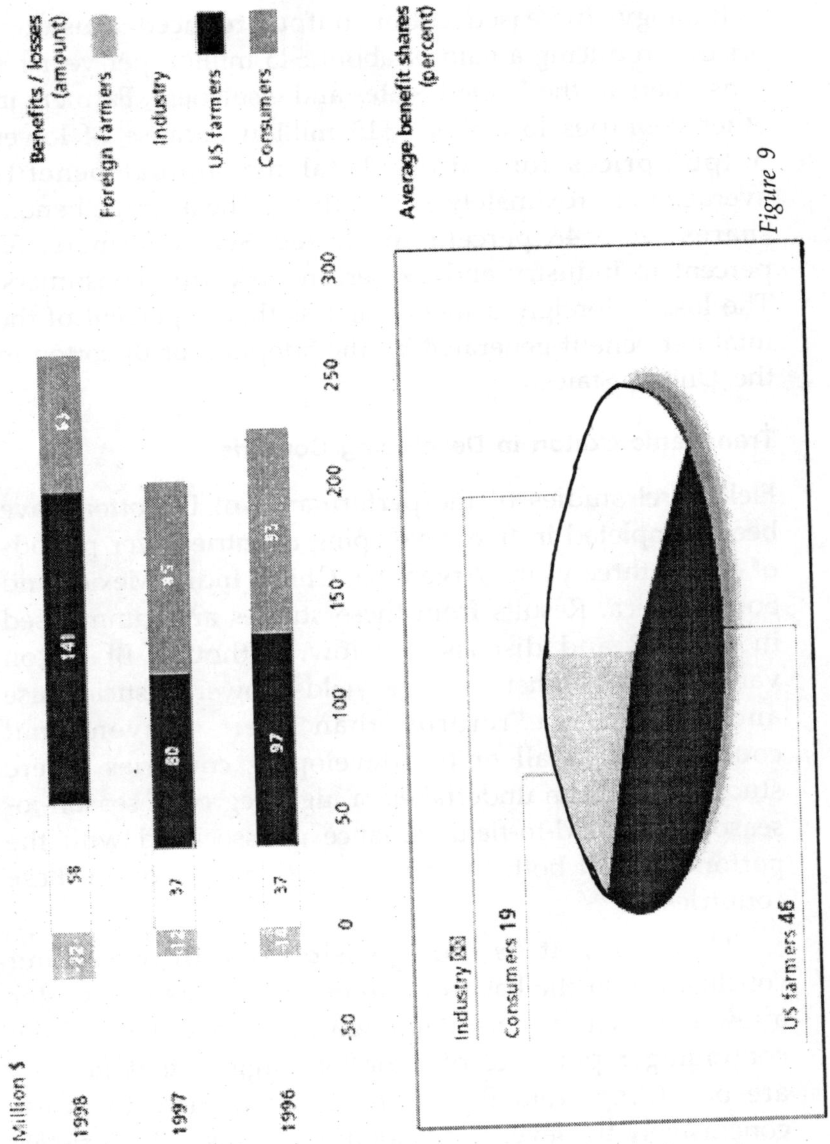
Million $
1998
1997
1996
-50
0
50
100
150
200
250
300
58
141
63
37
80
85
37
97
93
Benefits / losses (amount)
Foreign farmers
Industry
US farmers
Consumers
Average benefit shares (percent)
Industry
Consumers 19
US farmers 46

Figure 9

raised effective yields. The industry—primarily Monsanto and D&PL—earned about US$80 million from sales of Bt technology. Increased cotton output reduced consumer prices, producing a gain of about $45 million per year for consumers in the United States and elsewhere. Farmers in other countries lost about $15 million because of lower output prices for cotton. Total net annual benefits averaged approximately $215 million. The average benefit shares were 46 percent to United States farmers, 35 percent to industry and 19 percent to cotton consumers. The loss to foreign farmers was less than 1 percent of the total net benefit generated by the adoption of Bt cotton in the United States.

Transgenic Cotton in Developing Countries

Field-level studies of the performance of Bt cotton have been completed in five developing countries over periods of one to three years: Argentina, China, India, Mexico and South Africa. Results from these studies are summarised in Table 3 and discussed below. Although Bt cotton varieties had higher average yields, lower pesticide use and higher net returns than their conventional counterparts in all of the developing countries where studies have been undertaken, a high degree of season-to-season and field-to-field variance is associated with the performance of both Bt and conventional cotton in these countries.

Therefore, it is not possible to draw strong conclusions on the basis of two or three years of data for a few hundred farmers. Although the data so far and the continuing rapid pace of adoption suggest that farmers are benefiting from Bt cotton, it is too early to assess conclusively the level and stability of yields of Bt varieties compared with conventional varieties because these depend, among other factors, on pest infestations and agronomic practices, which vary widely.

Table 3. Performance differences between Bt and conventional cotton

	Argentina	China	India	Mexico	South Africa
Lint Yield					
(kg/ha)	531	523	699	165	237
(Percentage)	33	19	80	11	65
Cemical Sprays (no.)	-2.4	...	-3.0	-2.2	...
Gross Revenue					
($/ha)	121	262	...	248	59
(Percentage)	34	23	...	9	65
Pest Control					
($/ha)	-18	-230	-30	-106	-26
(Percentage)	-47	-67	...	-77	-58
Seed Costs					
($/ha)	87	32	...	58	14
(Percentage)	530	95	...	165	89
Total Costs					
($/ha)	99	-208	...	-47	2
(Percentage)	35	-16	...	-27	3
Profit					
($/ha)	23	470	...	295	65
(Percentage)	31	340	...	12	299

The distributional impacts of Bt cotton have been studied for Argentina, China, Mexico and South Africa. The available evidence indicates that transgenic cotton varieties are scale neutral with regard to both speed of adoption and per hectare benefits. In other words, small farmers are equally or more likely to benefit from Bt cotton as are larger farmers. This is not surprising given the manner in which Bt cotton varieties simplify the farmers' management task. Qaim and Zilberman argue that the relative performance of Bt cotton is likely to be greatest when used by small farmers in developing countries where pest pressure is high and access to effective chemical pest control is low, because of the large pest losses typically suffered by these farmers. This notion is supported by the international data available to date, which show the yield advantage to be largest in Argentina, China and India.

Argentina

Bt cotton in Argentina over two growing seasons, 1999/2000 and 2000/01. Bt cotton was first released in Argentina in 1998 by CDM Mandiyú SRL, a private joint venture between Monsanto, the Delta and Pine Land Company (D&PL) and the Argentine company Ciagro. The Bt varieties commercialised in Argentina were originally developed for the United States market. Bt cotton technology is patented in Argentina and farmers are required to pay technology fees. Under Argentine law, farmers are allowed to save and reproduce seed for one season before they are required to buy fresh certified material. However, Mandiyú requires farmers to sign special purchase contracts that prohibit the use of farm-saved seeds for Bt cotton. Unlike in other countries (or in the case of HT soybean in Argentina), the adoption of Bt cotton in Argentina has been slow and by 2001 had reached only about 5 percent of the total cotton area.

The yields for Bt cotton in Argentina averaged 531 kg/ha (or 33 percent) higher than for conventional varieties. Qaim and de Janvry note that the conventional varieties grown in Argentina are actually better adapted for local conditions and have higher agronomic potential yields than the Bt varieties, so the yield differential attributable to the reduction in pest damage to the Bt varieties would be even more than 33 percent. As there was little difference in market prices for Bt and non-Bt cotton, higher yields for the Bt varieties led to an average 34 percent increase in gross revenues.

The number of pesticide applications was lower and pesticide costs were reduced almost by half. Seed costs, however, were more than six times higher for the Bt varieties than for conventional varieties and, as a result, total variable costs were 35 percent higher. Net revenues were higher for Bt than for non-Bt varieties, but by a fairly small absolute value and by a significantly smaller margin than in other countries.

Qaim and de Janvry conclude that high seed costs are the primary reason for the relatively low farm-level profit margins for Bt cotton in Argentina, which in turn explains the low rate of Bt cotton adoption compared with the rapid adoption of HT soybeans in that country. They use a contingent valuation method to estimate that the price Argentine farmers would be willing to pay for Bt seeds is less than half of the actual price. At this lower price, farmers' net returns would significantly increase, but company revenues would also rise because farmers would buy more seed. This finding raises an important question regarding why Mandiyú would charge prices higher than their profit-maximising level. It also raises concerns regarding the long-term potential for private monopolies to extract excess profits from farmers in the absence of competition or appropriate regulatory constraints on monopoly power.

Genetically engineered Herbicide-tolerant crops feature a gene from the soil bacterium Agrobacterium tumefaciens, which makes the recipient plant tolerant to the broad-spectrum herbicide glyphosate. Introduced to a crop plant, the technology can facilitate weed management in farmers' fields. It can reduce production costs, through the substitution of glyphosate for an array of more expensive (and more toxic) herbicides. The timing and choice of herbicide is simplified for HT crops because glyphosate effectively controls both broad-leaved weeds and grasses and has a fairly broad window for the timing of application.

Herbicide tolerance for various crops was developed by Monsanto under the name Roundup Ready©(RR). RR soybeans were commercially released in Argentina and the United States in 1996. The sale and use of RR technology is protected in the United States through patents and a sales contract with farmers, but neither form of intellectual property protection is used in Argentina. Thus, in Argentina, RR soybeans are widely available from sources other than Monsanto and Argentine farmers are legally allowed to use farm-saved seeds. As a result, Argentine farmers pay a relatively small price premium for RR of about 30 percent, whereas farmers in the United States on average pay 43 percent more. Adoption proceeded rapidly in both countries.

By 2002, an estimated 99 percent of the Argentine soybean area and 75 percent of the United States area were cultivated with RR seeds. Yields of RR soybeans are not significantly different from yields of conventional soybeans in either Argentina or the United States, but reduced herbicide and tillage costs generate farm-level benefits. Many farmers switched to low-till or even no-till cultivation practices after the adoption of RR soybeans, reducing machinery and labour costs and improving soil conservation. Harvesting costs are also lower because of the lower incidence of green weeds.

In Argentina, the total variable cost of production is about 8 percent ($21/ha) lower for RR soybeans than for a conventional crop. Results for the United States are less clear. Moschini, Lapan and Sobolevsky (2000) estimated a cost advantage of $20/ha for 2000 for the United States as a whole, and Duffy (2001) found negligible cost savings in Iowa in 1998 and 2000. Taking an average over all sources, it appears that cost savings in the United States are similar to those in Argentina.Qaim and Traxler estimated that RR soybeans created more than $1.2 billion in economic benefits in 2001, about 4 percent of the value of the world soybean crop.

Soybean consumers worldwide gained $652 million (53 percent of total benefits) as a result of lower prices. Seed firms received $421 million (34 percent) as technology revenue,1 most of which came from the United States market. Soybean producers in Argentina and the United States received benefits of more than $300 million and $145 million, respectively, whereas producers in countries where RR technology is not available faced losses of $291 million in 2001 as a result of the induced decline of about 2 percent ($4.06 per tonne) in world market prices.

Farmers as a group received a net benefit of $158 million, 13 percent of total economic gains produced by the technology. As in the cotton studies, gross technology revenues are used as a measure of monopoly rent. No research, marketing or administration costs are deducted. If we assume, for example, that these costs amount to 33 percent of technology fee revenues, the monopoly rent would fall to around $280 million (26 percent of total surplus).

China

More than 4 million small farmers in China are growing Bt cotton on about 30 percent of China's total cotton area. China's share of global Bt cotton area has increased dramatically since it was first commercialised in 1997 to more than 35 percent in 2001. Pray et al. surveyed cotton farmers in China over three seasons from 1999 to 2001.

The surveys were conducted in the main cotton-growing provinces where both Bt and non-Bt varieties were available. The initial survey included farmers in Hebei and Shandong Provinces. Adoption has advanced rapidly in these provinces because bollworms are the major pest and severe resistance to chemical pesticides is widespread.

Adoption approaches 100 percent in Hebei and exceeds 80 percent in Shandong. Henan Province was added to the survey in 2000. Bt adoption has levelled off at about 30 percent in Henan despite heavy pressure from bollworms, reportedly because farmers there do not have access to the best Bt varieties. Anhui and Jiangsu Provinces were added to the study in 2001. Adoption started later and has been slower in these provinces partly because red spider mites (which are not susceptible to Bt) are a more serious problem there.

For China, the yield advantage for Bt cotton averaged 523 kg/ha or 19 percent compared with conventional varieties over the three-year period from 1999 to 2001. This translated into an average revenue gain of 23 percent. Seed costs for the Bt varieties were almost double those for conventional varieties. Compared with the Argentine case, however, this price premium is quite low. Pray et al. attribute the relatively low price premium for Bt seed to the presence of strong competition in the market between the CAAS varieties developed by the public sector and those available from Monsanto.

Offsetting the seed price premium, pesticide costs were 67 percent lower, and total costs were 16 percent lower than for conventional cotton. Total profits averaged $470 more per hectare for the Bt producers than for the non-Bt producers, who in fact lost money in each of the three years. Pray et al. estimate that Bt cotton farmers in China reduced their use of chemical pesticides by an average of 43.8kg/ha compared with conventional cotton farmers. The largest reductions were in Hebei and Shandong Provinces, where bollworms are the major pest. Lower pesticide use translated into lower costs for chemicals and labour for spraying, but additional environmental and health benefits were also found.

Table 4. Distribution of benefits of Bt cotton adoption by size of farm or income class in China, 1999

		(kg/ha)	($/ha)	($/ha)
	Bt as percentage of observations	Yield increase	Change in total cost	Change in net income
Farm Size				
0.0-0.47 ha	86	410	-162	401
0.47-1 ha	85	-134	-534	466
1+ ha	87	-124	-182	185
Houshold Income ($)				
1-1 200	85	170	-302	380
1 200+	91	65	-54	157
Per Capit Income ($)				
1-180	85	456	-215	446
180-360	83	8	-284	303
360+	97	-60	1	-15

As a result of Bt cotton, pesticide use in China was reduced by an estimated 78 000 tonnes in 2001, an amount equal to about one-quarter of the total quantity of chemical pesticides used in China in a typical year. Because chemicals are typically applied with backpack sprayers in China and farmers rarely use protective clothing, they are often exposed to dangerous levels of pesticide. Bt cotton farmers experienced a much lower incidence of pesticide poisonings than those growing conventional varieties (5-8 percent vs 12-29 percent).

Pray and Huang looked at the distribution of economic benefits in China by farm size and income class. They found that farms of less than 1 ha had more than double the net increase in per hectare income of those larger than 1 ha (Table 4). Poorer households and individuals also received a much larger per hectare increase in net incomes than richer ones. These results suggest that Bt cotton is generating large pro-poor gains in net income in China.

India

Bt cotton was only approved for commercialisation in India in 2003 and therefore market-based studies are not yet available. Qaim and Zilberman analysed Indian field trial data from 2001 and reported changes in crop yields and pesticide use between conventional and Bt cotton. The trials were initiated by the Indian company Maharashtra Hybrid Seed Company on 395 farms in seven Indian states. The trials were supervised by regulatory authorities and managed by farmers using customary practices. The study compared yield performance and chemical use for a Bt hybrid, the same hybrid without the Bt gene, and a popular non-Bt variety grown on adjacent 646 m2 plots. The analysis was based on results from 157 representative farms for which comprehensive records were kept. Table 3 on page 48 reports the comparison between the Bt hybrid and the same hybrid without the Bt gene.

Average effective yields for the Bt hybrid exceeded those for the non-Bt hybrid by 80 percent, reflecting high levels of pest pressure during the growing season and a lack of alternative pest control options. This yield differential is much higher than that found in China, Mexico and the United States. Qaim and Zilberman argue that the performance differential for Bt cotton is higher in

India than elsewhere because pest pressure is high and farmers do not have access to affordable and effective pesticides. They argue further that the non-Bt hybrid and popular varieties had similarly poor performance, suggesting that yield potential was not a factor in the performance differential between the Bt and non-Bt hybrids.

Qaim and Zilberman report that insecticide resistance is widespread in India, so that ever-increasing amounts of pesticide have to be sprayed each year. Their survey results for 2001 showed the number of chemical sprays against bollworms was reduced from an average of 3.68 to 0.62 per season, although the number of sprays against other insects was not significantly different. The overall amount of insecticide use was reduced by 69 percent, with almost all the reduction occurring in highly hazardous organophosphates, carbamates and pyrethroids belonging to international toxicity classes I and II.

Mexico

The amount of cotton planted in Mexico varies widely from year to year depending on government policies, exchange rates, world prices and—critically—the availability of water for irrigation. Cotton area declined from about 250 000 ha in the mid-1990s to about 80 000 ha in 2000, whereas the share planted to Bt varieties grew from about 5 percent to 33 percent. Bt adoption patterns in Mexico reflect regional patterns of pest infestation and economic losses resulting from pest damage. Adoption has been most rapid in Comarca Lagunera, a region that comprises parts of the states of Coahuila and Durango, and the region most critically affected by bollworms. The other cotton-growing regions of Mexico are afflicted with boll weevil and other pests that are not susceptible to Bt and thus require the use of chemical controls. Bt adoption is correspondingly low in these regions. Bt cotton is barred from the southern states of Chiapas and Yucatan where wild species of Gossypium hirsutum, a native relative of cotton, exist.

The Bt cotton varieties grown in Mexico were developed originally for the United States market by D&PL in cooperation with Monsanto. Monsanto requires farmers in Mexico to sign a seed contract that forbids them from saving seed and requires them to have their cotton ginned only at Monsanto-authorised mills. The contract also requires farmers to follow a specified resistance management strategy and to permit Monsanto agents to inspect their fields for compliance with refugia and seed-saving restrictions.

Cotton producers in Comarca Lagunera are generally classified as falling into one of three groups: ejidos, small landholders and independent producers. Ejidos have landholdings of 2-10 ha, small producers 30-40 ha and independent producers somewhat more but typically less than 100 ha. Ejidos and small landholders are organised into farmer associations for the purposes of obtaining credit and technical assistance. Each farmer group has a technical consultant who works for the association. Traxler et al. surveyed cotton farmers in Comarca Lagunera for the 1997 and 1998 growing seasons through the technical consultants working for the association SEREASA.

The association is one of the largest in Comarca Lagunera, and had 638 farmers owning almost 5,000 ha of land during the study period. Of this total area, between 2,000 and 2,500 ha were planted to cotton, about 12 percent of the cotton area in Comarca Lagunera. Bt varieties were planted on 52 percent of the cotton area in

Comarca Lagunera in 1997, increasing to 72 percent in 1998.

The average effective yield differential between Bt and conventional cotton was 165 kg/ha or about 11 percent, considerably lower than for the other countries shown in Table 3. The yield differential varied sharply over the two growing seasons covered by the survey, from almost nil in 1997 to 20 percent in 1998. 1997 was a year of very low pest pressure in Comarca Lagunera. Pesticide costs were about 77 percent lower for Bt than for conventional cotton, and the number of chemical sprays was lower. Seed costs were almost three times higher for Bt cotton, reflecting a fairly high technology premium. As a result, the average profit differential for the two years was $295/ha. This varied from less than $8 in 1997 to $582 in 1998.

Traxler et al. calculated the distribution of the economic benefits from Bt cotton in Comarca Lagunera between the farmers in the region and the companies supplying the Bt varieties, Monsanto and D&PL. For the two years of the study, farmers captured an average of 86 percent of the total benefits, compared with 14 percent for the germplasm suppliers. The per hectare change in profit accruing to farmers varied widely between the two years, as noted above. As a result, the total producer surplus ranged from less than $35 000 to almost $5 million. For the two years, an estimated total of almost $5.5 million in benefits was produced, most of it in the second year and most of it captured by farmers.

In this calculation the entire amount attributed to Monsanto and D&PL cannot be considered truly a net benefit to the companies, because costs such as seed distribution, administration and marketing costs were not accounted for. A revenue of $1.5 million from seed sales is not a large sum for a company like Monsanto, which has $5.49 billion in annual revenue. The large annual fluctuations are largely caused by variability in pest infestation levels; in years of heavy pest pressure, Bt cotton produces a large advantage over conventional cotton varieties. Because Mexico grows a small share of the world's cotton, there are no economy-wide effects on prices or consumer welfare.

South Africa

Bt cotton was the first transgenic crop to be commercially released in sub-Saharan Africa following the implementation in 1999 of the Genetically Modified Organisms Act, 1997. By 2002 some 30,000 ha of Bt cotton were planted in South Africa, of which about 5,700 ha were in the Makhathini Flats area of KwaZulu-Natal Province. Bennett, Morse and Ismael examined the experience of resource-poor smallholder cotton farmers in the Makhathini Flats.

Vunisa Cotton is a private commercial company in the Makhathini Flats that supplies farmers with cotton inputs (seed, pesticide and credit) and buys their output. Bennett, Morse and Ismael used individual farmer records held by Vunisa Cotton to collect information on input use, yields, farm characteristics and other information for the three growing seasons beginning in 1998/99. In addition, personal interviews were undertaken with a random sample of smallholder farmers in 1998/99 and 1999/2000, and 32 in-depth case study interviews were conducted in 2000/01.

Bt cotton benefited from higher yields (as a result of less pest damage), lower pesticide use and less labour for pesticide applications. Yields were an average 264 kg/ha (65 percent) higher for the adopters. The yield differential was particularly large in the poor, wet growing season of

1999/2000, reaching 85 percent. Adopters used less seed per hectare than non-adopters, but higher prices for Bt seed meant that total seed costs were 89 percent higher. This was offset by lower pesticide and labour costs, so total costs were only 3 percent higher for Bt cotton on average. Higher yields and nearly equal costs meant that Bt adopters achieved net profits 3-4 times higher than those of conventional producers in all growing seasons, with the differential being especially large in 1999/2000, when conventional growers lost money.

In 1997/98, Vunisa Cotton purposely targeted the release of Bt cotton to a few, relatively large, farmers. By 1998/99, the first growing season of this study, approximately 10 percent of smallholders in Makhathini had adopted Bt cotton, followed by 25 percent the second year and 50 percent the third year. By the fourth season, 2001/02, which was not covered in the analysis because of data limitations, an estimated 92 percent of smallholder cotton farmers in the region had adopted the Bt variety. Older, male and wealthier farmers were more likely to adopt in the first season, but by the second and third seasons, smaller farmers of various ages and both genders were also growing Bt cotton. Their analysis showed that smaller farmers growing Bt cotton actually earned higher per hectare gross margins than did larger Bt cotton growers.

In a study of five West African cotton-producing countries, Cabanilla, Abdoulaye and Sanders examined the economic benefits that could accrue to cotton farmers if Bt cotton were introduced to the region. Cotton is a major source of export revenue in these countries—Benin, Burkina Faso, Côte d'Ivoire, Mali and Senegal—and a source of cash income for millions of resource-poor farmers. Depending on the rate of adoption and the actual yield advantage, the potential benefits for these countries as a group could range from $21 million to $205 million.

Cabanilla, Abdoulaye and Sanders based their analysis on the similarities between pest populations and chemical use in these countries with those found in other developing countries where Bt cotton has been introduced. The major insect pests in West Africa are bollworms, which are currently controlled by spraying up to seven times per season with broad-spectrum insecticides, usually a combination of organophosphates and pyrethroids.

9

Genetic Erosion of Agricultural Crops

Global Biodiversity Assessment issued by the United Nations Environment Programme (UNEP) in 1995 that the loss of diversity of agricultural crops "threatens humanity's food supplies." More generally, there is increasing awareness of the substantial loss of genetic diversity of crops and the negative impacts of this on poorer farmers and, potentially, on the future of agricultural development.

The general magnitude of genetic erosion that has occurred is suggested by several examples. For example, a study by the Rural Advancement Fund International (RAFI) that compared the number of varieties of different commercial crops known to the U. S. Department of Agriculture in 1903 to the number of varieties of these crops for which seeds existed in the National Seed Storage Laboratory (NSSL) in 1983. Considering about 75 different vegetables together, the RAFI study found that approximately 97% of the varieties on the 1903 lists are now extinct. Among selected fruits, the RAFI study found that 6,121 out of 7,098 (86.2%) varieties of apple are now extinct as are 2,354 out of 2,683 (87.7%) varieties of pear.

Examples of the losses for specific crops are presented in Table 1. Or, as another example, in the late 19th century about 50 native varieties of wheat were known in Sicily; more recent studies estimate that no more 20 of these remain.

Table 1. U. S. Vegetable Varieties Lost, 1903-1983

Vegetable Common Name	Vegetable Scientific Name	Total Varieties 1903	Varieties in U.S. NSSL[1] Collection, 1983	Varieties Lost(%)
Asparagus	Asparagus officinalis	46	1	45(97.8)
Lima Bean	Phaseolus lunatus	96	8	88(91.7)
Brussels Sprouts	Brassica oleracea, var. gemmifera	35	4	31(88.6)
Carrot	Daucus carota	287	21	266(92.7)
Cauliflower	Brassica oleracea, var. botrytis	158	9	149(94.3)
Celery	Apium graveolens, var. dulce	164	3	161(98.2)
Sweet Corn	Zea mays	307	12	295(96.10
Gherkin	Cucumis anguria	10	2	8(80)
Eggplant	Solanum melongena	97	9	88(90.7)
Kale	Brassica oleracea, var. acephala	124	9	115(92.7)
Leek	Allium ampelo-prasum	39	5	34(87.2)
Pea	Pisum sativum	408	25	383(93.9)
Peppers	Capsicum annuum	126	13	113(89.7)
Squash	Cucurbita spp.	341	40	301(88.3)
Tomato	Lycopersicon esculentum	408	79	329(80.6)
Watermelon (and citron)	Citrullus lanatus	223	20	203(91.0)

[1]: NSSL = National Seed Storage Laboratory

Genetic erosion also occurs among indigenous groups in the United States. One study found that the Hopi Indians, who live in a semi-arid area of northern Arizona, have lost some traditional blue corn varieties due to voluntary species replacement, decreases in traditional seed saving skills, and to changes in crops' vulnerabilities to pests, weeds, etc.

Genetic erosion is not limited to the United States or other developed countries. In China, of almost 10,000 varieties of wheat in use in 1949, only about 1000 remained by the 1970s. In Thailand, wild rice populations grow in ditches along roadways; during the period from 1983-1991 roadway expansions and factory construction resulted in the loss of several wild rice populations. The loss of these localised populations has led to a dramatic reduction of wild rice varieties at sites near the disturbances.

The underlying basis for the claims of impending doom if genetic diversity continues to be lost is the recognition that as the genetic base of crops narrows, crop vulnerability to pests, disease, and atypical climate conditions increases. As a greater proportion of cropland is planted in fewer and fewer varieties of a given crop, the potential for a major failure increases. Recent history is replete with examples of major crop failures attributed to these causes, with significant humanitarian and economic consequences.

Perhaps the most famous catastrophe was the Irish Potato Famine of the 1840s. It has been attributed to potato blight Phytophtora infestans, to which the genetically similar Irish potatoes were susceptible. More recent crop failures with similar causes have occurred in the United States and in the (former) Soviet Union. Large-scale crop losses due to susceptibility to a single pest occurred in the United States in the 1940s when 80% of the U.S. oat crop was eliminated. Another large loss occurred in the 1970s when corn blight wiped out a large part of the United States' corn crop.

And, in the (former) Soviet Union, lulled by warmer than normal winters, farmers planted a high yielding wheat variety in more northern latitudes, only to have a disaster in 1972 when a severe winter caused a large shortfall in the winter wheat crop. Although these disasters were caused by lack of diversity in large croplands, it is not clear that genetic erosion, i.e., the loss of genetic varieties, is itself the cause. However, it does serve to underscore the absolute necessity to avoid genetic erosion.

When disaster strikes, the solution is usually to turn to other varieties of the same crop that are resistant to the cause of the original problem. For example, in the 1970s, when the Indian rice crop was devastated by the Tungro virus, scientists had to analyse 17,000 varieties of rice before finding a single wild variety that was resistant. What might the consequences have been if this variety had been lost? Have Indian farmers and, more generally, the world's farmers learned the lessons of over-reliance on a narrow set of crop varieties? Apparently not; of the more than 30,000 varieties of rice once cultivated in India, today over 75% of India's rice is limited to 1 of 10 varieties, while in the Philippines over 90% of the rice sown is limited to two varieties as is the maize in Zimbabwe. Clearly, such behaviour courts disaster.

The value maintaining diversity within crop varieties was reinforced by the United States' experience with a certain wheat variety, collected by Jack Harlan in Turkey in 1948. Because it was tall, thin stemmed and susceptible to leaf rust and cold weather, the variety apparently had little to commend it. However, it saved the day when stripe rust became a problem in the northwestern United States. Harlan's wheat variety was found to be resistant to four types of stripe rust, ten types of dwarf bunt, and to have good tolerance to flag smut and snow mould. Consequently, this wheat variety is now used in all wheat breeding programmes in the northwestern United States.

In some cases the loss of particular crop varieties is not complete, but instead reduces surviving members of a landrace to a few isolated populations. In such cases there is significant risk of the ultimate loss of diversity, because small populations will lead to increased inbreeding which reduces the fitness of individual plants and hence may lead to extinction. The above examples show that while there may economic advantages in planting large areas to a single variety of a crop, it is prudent to have the widest possible genetic base for use in improving the resistance of crops to sources of failure. This in turn leads to the conclusion that the disappearance of crop varieties by whatever means and for whatever reason poses a threat of unknown but potentially great magnitude to agriculture.

Causes of Genetic Erosion

The primary causes of genetic erosion have been identified by the United Nations Food and Agriculture Organisation (FAO) as "urban expansion, agricultural modernisation, civil strife and war". For example, Zimmerer includes changes in the availability of labour in peasant household economies, changes in access to land and capital, dietary importance and "culinary esteem" of different crops, and variation in the spatial patterning of crop varieties as important determinants in the loss or maintenance of different cultivars. Arunachalam adds to this list loss of habitats due to such natural disasters as floods, earthquakes, soil erosion, etc., and loss of germplasm during storage.

It is important to note that wild varieties are also considered when discussing genetic erosion; recent advances in plant breeding make it possible to use strains from more distant genetic relatives in cultivar development than was formerly possible. Therefore, the loss of more distantly related varieties of a crop may ultimately limit breeding options. Perhaps the most significant cause of genetic erosion is the introduction and growing use of modern varieties of agricultural crops.

The importance of this as a cause of genetic erosion stems from the consequent dropping of wild varieties and traditional local varieties in favour of the higher yielding, more uniform individual plants possible with modern varieties. As farmers see increases in yields (and in economic returns), they increasingly replace traditional varieties with modern varieties in many fields. There is also an economic incentive to plant additional, new fields with these varieties, thereby displacing wild varieties and hybrids that may be growing unattended.

Another prominent cause of genetic erosion is land use change. The example cited earlier, of the loss of some populations of wild rice due to roadway expansion in Thailand, illustrates the problem on a very small scale. On larger scales, not only may populations of wild varieties be lost to urbanisation or other land use change, the processes themselves may spawn various types of pollution that can affect environmental conditions (and thus wild crop populations) far from the original site. An example of "off site impact" is the Freeport-McMoran mine in Irian Jaya, Indonesia.

Operations at this large gold and copper mine have caused the destruction of thousands of acres tropical rainforest at elevations over 13,500 feet. This was accompanied by the eviction of over 5000 tribal people who were living in the mine area, and the daily dumping of over 120,000 tons of mine waste into the Ajakwa River System (with planned expansion to 300,000 tons daily in the next few years). Secondary impacts include high levels of contamination of river water due to heavy metal concentrates with consequential loss of native fish in the rivers, and the Ajakwa river having breached its banks and deposited contaminated water over a large floodplain.

While there has been no known analysis of possible genetic erosion of agricultural crop varieties in this area, it is entirely possible due to large-scale changes in population, land use, economy, and environmental degradation. Possible effects are the loss of some crop varieties existing in small wild populations, or cultivated crop varieties existing only in the gardens of evicted people. Additionally, the failure of the evicted people to bring seeds of traditional crop varieties to their new homes, poor performance of old varieties in the new home areas, and the possible destruction of highly localised crop varieties by the polluted river water may all have contributed to genetic erosion.

Another potential source of genetic erosion of agricultural crops is change in lifestyle of the peasant populations who have developed or maintain them. Zimmerer found that change (reduction) in the available labour supply in a peasant economy can result in abandonment of crop varieties requiring high amounts of labour. More typically, farmers may choose modern over traditional varieties that have higher returns for a given labour input. Alternately, if a large, new employment opportunity develops, people may prefer to put their efforts into wage labour than in traditional agriculture.

Ironically, the Freeport McMaron mine in Irian Jaya may also contribute to genetic erosion by offering employment to local people. Finally, it is possible that in some sub-Saharan countries with very high rates of people who are HIV positive or who have AIDS, as people become sick and the available labour pool shrinks, labour-intensive crop varieties may no longer be cultivated. Some of the secondary effects of war and civil unrest have been known to induce genetic erosion as a byproduct. There are several processes by which this can occur. First, combatants may take military action to stop farmers from growing food for the "other side." This has occurred in Sudan, which experienced famine as the result of a civil war in which both sides kept farmers from the fields and otherwise disrupted access to food.

Also, farmers may flee combat areas, thereby leaving fields with local crop varieties untended; in the case of varieties that are labour intensive, yields may be quite low or may fail altogether. Farmers may also be unable to work their fields where land mines have been seeded. For example, substantial land mine seeding in Mozambique along that country's border with Malawi kept some farmers from their fields for years; in cases where unique, localised varieties were grown, this may have resulted in some loss of genetic diversity.

Finally, farmers who experience food shortages as a result of war disrupting food production may eat the seeds of crop varieties to stave off starvation. This has been reported from Somalia, in which the country's seed supplies were raided and consumed. The only factor that prevented a loss of genetic diversity was the Somalian seed stocks stored in Kenya, India, Nigeria, and elsewhere. This last example points to a conclusion reached after the Rwanda civil war about genetic erosion as a consequence of war: sometimes genetic erosion may be governed more by what happens to the seed supplies and channels for resupply than to local seed stocks and plants.

Natural disasters can also bring about genetic erosion. Floods can eradicate unique varieties by inundating and drowning local populations. At the other extreme, drought can also cause some loss of genetic diversity; droughts of just a single season can result in people consuming seed stocks, while successive years of drought can prompt changes in cropping patterns and the geographic distribution of crops. These changes in cropping patterns may also include the use of alternative, more drought resistant crops in preference to the traditional landraces.

A Framework for World Information and Early Warning System

The recognised need to prevent genetic erosion of agricultural crops has led to various initiatives. One of these, the United Nations Food and Agricultural Organisation (FAO) World Information and Early Warning System on Plant Genetic Resources (WIEWS), is intended to provide information on national programmes of plant genetic resources for food and agriculture (PGRFA). This system consists of a number of data bases and information retrieval software for both in situ and ex situ losses; the former refers to losses in the field, while the latter to losses in seedbanks and other germplasm collections.

One of the WIEWS databases, the "Early Warning" database, is under development and will consist of reports of possible genetic erosion in both the natural environments and in ex situ collections. In fact, three report formats have been defined; these cover genetic erosion during storage, genetic erosion of wild crop relatives, and genetic erosion of local varieties. In spite of its name, WIEWS does not, by its own admission, currently have a system to provide an early warning of potential losses. In their own words, they do not have "a bell to ring".

Detecting and assessing in situ genetic erosion has been suggested as the first priority in any major effort to arrest loss of genetic diversity. For that reason, and also because the causes of ex situ losses (ineffective storage of germplasm) and the detection and assessment of these losses would be based on completely different conceptual models. As noted, WIEWS is not operational. The whole issue of targeting plants for conservation efforts and the criteria that might be used to select them is relatively new and the proposals are quite general.

In terms of the applications of early warning systems, we are familiar only with famine early warning which was initiated as a formal activity by the Global Information and Early Warning System (GIEWS) of FAO in 1975. Since that time, a handful of famine and food security early warning systems have been developed, each serving a slightly different audiences and using different methods. Because of its comparatively longer history, we feel it may also provide some insight in terms of the types of methods that might be employed by WIEWS.

Evolution in Early Warning Methodologies

Since introduced a quarter century ago, there has been a gradual evolution in early warning methodologies. Initially, the famine early warning task was perceived as simply monitoring aggregate data to establish "food balance" (i.e., identification of areas of food surplus and deficit). However, this approach focuses only on the availability of food and does not recognise that regions and groups have differential access to food (e.g., rich households have better access than poor ones). To compensate for this weakness (as well as problems with the quality of agricultural production data) more focused systems came to rely on a "convergence of evidence"—general agreement among a set of independent indicators. For example, if a deficit was forecast for an area and it experienced an extreme increase in cereal prices, it was judged to be at risk.

Subsequent development of early warning systems has proceeded down two evolutionary paths. The first is a continued refinement of the methods described above, or a deductive approach (top down). It relies on wider array of generally available secondary data that is usually aggregated (e.g., census or agricultural production data reported by administrative unit), or relatively low spatial resolution satellite imagery. In addition, it is obvious that the likelihood of an emergency occurring and the potential severity of such an emergency (i.e., the level of vulnerability to food insecurity) are unevenly distributed among regions and groups.

Thus, as a move toward more efficiency and accuracy, a socioeconomic vulnerability assessment (based on secondary data) was added to identify the most vulnerable regions and groups within the regions of interest. This allows attention to be targeted on the most "vulnerable" regions. Additionally, it permits development of generic models of household behaviour that help to interpret and predict the behaviour of monitored data and their impacts (e.g., cereal prices increase and livestock prices decline when crops fail and such price increases have different effects on different classes of households).

Because it relies on secondary data, the chief advantage of the deductive approach is efficiency in terms of time and cost. The primary disadvantage is the risk of making incorrect inferences about actual conditions on the ground (i.e., assuming that conditions are good/bad for specific classes of households) from highly aggregated data. The second inductive approach to early warning, addresses this fundamental shortcoming by developing specific models of household behaviour based on field interviews.

Using these models, the impact of a specific event (e.g., 30% decline in maize yields due to drought) can be estimated for each class of households. Thus, a 30% decline in "income" for subsistence farmers relying totally on maize may be substantially greater than for farmers relying on a mix of maize and livestock. The advantage of this approach is that it captures the true nature of household economies. The main disadvantages are the cost of fielding survey teams, and the considerable error in food aid estimates that are accumulated by extending the models over large areas.

Originally, the inductive and deductive approaches to famine early warning were viewed as competitive. However, with time, they have come to be viewed as complementary in terms of their relative strengths and steps are being taken to merge them into a single, unified approach .

Development of WIEWS

Experience with the famine early warning system provides some guidance. However, given the different natures of the problems steps that are taken to redress them, it is clear that the famine early warning experience does not necessarily provide a template.

Requirements System

Following the general famine early warning model, it is necessary to specify the information requirements that the system is intended to satisfy. These can be framed through a set of questions:

1. *Early warning of what*? Before embarking on development of a system, the conditions that it is intended to identify must be clearly defined.
2. *Early warning for what*? Early warning information will be used to initiate mitigation. The types of action that might be taken to address the problem will help define the performance parameters of the system.
3. *How early is early enough*? If we assume that this is a global system, deciding how soon a warning is necessary is of critical concern (e.g., must assessments be performed continuously, or only every 3 - 5 years? How long a "lead time" is required to mobilise mitigation?).
4. *How well do we need to know it*? Issues of accuracy and scale are driven by how the data are intended to be used (e.g., is it better to err on the side of false warnings to ensure that none are missed?) and the level of effort that is envisioned (i.e., budget).

Caveats

The actions that an early warning system are intended to initiate must be defined before the system can be developed. The international and large binational famine early warning systems are intended to provide information that will mobilise large efforts to provide food assistance. It is possible that WIEWS is also intended to initiate such large-scale actions, such as campaigns to locate, collect and store germplasm from threatened areas. However, it is equally likely that something quite different might be pursued (i.e., national conservation campaigns; in situ conservation programmes) .

Vulnerability Assessment for Famine

The famine early warning system model can be applied to any area in which either the perceived risk or the potential implications of genetic erosion are considered to be unacceptable. While, in theory the system can be employed on a worldwide basis, it is both more feasible and more practical to apply it to select areas, or to encourage national initiatives using a common set of methods. As in the vulnerability assessments for famine, we may assume that, in considering genetic erosion, some areas are more susceptible or important than others. As noted in Heywood and Davis, "there are convincing scientific, economic, and sociological reasons for giving priority to the major centers of plant diversity throughout the world."

The conceptual model proposed here is similar to that used for many food security and vulnerability analyses. The deductive model described above is often referred to as the "indicator approach." It is based on the measurement and cumulative assessment of key applicable parameters (indicators) that measure or reflect conditions believed to be precursors of the conditions for which early warning is desired. In this context, the model calls for the collection and cumulative evaluation of parameters that measure the impending risk of genetic erosion to agricultural crop varieties.

Remote Sensing

The general types of information might be gathered that is pertinent to monitoring genetic erosion. Some of these include remote sensing. Obviously, as the list of potential causes and indicators that might be monitored is revised, the role of remote sensing may change. Perhaps more importantly, as the scale of assessment varies, so too may the role remote sensing plays. For example, if changes in regional land cover/land use over a 25-year period are critical indicators, the Landsat satellite record provides the only means of providing that information. Similarly, if there is a need to understand regional changes in farm-level cropping practices, sample aerial photography—or perhaps commercial high-resolution satellite imagery—may be the only data source that can meet that requirement. In addition, there a large number of other types of satellite (e.g., Radarsat) that may provide special information in other situations (e.g., areas with persistent cloud cover; areas with subtle changes in vegetation types).

The potential value of remote sensing data cannot be realised, however, if it is not firmly linked to conditions on the ground. This can be done in several ways, and almost always within a geographic information systems (GIS) environment that makes all observations spatially explicit. The first is through the use of a convergence of indicators: combining remote sensing data with other independent observations (e.g., growing season outlook from satellite data combined with yield forecasts developed on the ground). The second, and more expensive, alternative is an orchestrated campaign of ground sampling in support of remote sensing data acquisition. The third is to incorporate remote sensing data as in input into predictive models of behaviour. As noted, this last is the approach that is emerging now in the famine early warning arena.

An information system can be developed to provide early warning of the erosion of critical plant genetic resources. Technically, it is possible to envision an early warning system that would incorporate remote sensing and other geospatial information technologies (i.e., GIS, GPS) in support of conventional ground-based assessment methods that could efficiently provide accurate information. In practical terms, it is unlikely that such a system could be supported (1) if implemented through a large centralised initiative (as with the major famine early warning systems), or (2) if it were expected to provide global coverage on a routine basis. Moreover, it is highly unlikely that interventions could be carried out at anything other than a national level to arrest erosion of plant genetic resources.

10

Technology for Producing Healthy Seeds

Transmissible Plant Pathogens

Pathogens capable of causing systemic infections on their host plants can usually also be transmitted through vegetative propagation from infected mother plants to the young plants. If all the seedlings or planting materials are infected by this type of pathogen, there may be severe losses of the crop after planting. Infected seedlings or planting materials can easily be transported to new areas, resulting in the invasion and establishment of new diseases. In areas where the pathogens already exist, infected seedlings or planting materials serve as inoculum sources of pathogens that can be transmitted by insects and other vectors, causing outbreaks of disease.

Because these pathogens are always located inside the cells and systemically spread throughout the plant, the diseases they cause are often very difficult to control by conventional measures such as pesticide sprays. Experience over several decades tells us that the most effective way of dealing with this problem is to produce pathogen-free seedlings or planting materials.

In Taiwan, the first propagation system for virus-free planting materials was established in the early 1970s, for the mass production of virus-free seed potatoes. It has been in operation for more than 26 years, and has proved to be effective in controlling potato virus diseases. Since then, more than nine more crops have been included in pathogen-free propagation programmes. Before the 1980s, the most widely used pathogen detection techniques were antibody-based, such as enzyme-linked immuno-sorbent assay (ELISA). These techniques are still widely used in laboratories, and have also been adapted to be used on an industrial scale. The reason for their wide acceptance by industry lies in the fact that their sensitivity and specificity are at an acceptable and reasonable level. Furthermore, the results can be reproduced, while there is a minimum of false positive and negative results. In all, ELISA testing is efficient and cost effective.

However, the invention of DNA-based technologies in the 1980s has driven most research on pathogen detection to take a new direction. Numerous reports have shown that there are many more DNA-based pathogen detection techniques than serological tests such as ELISA. Nevertheless, most of the DNA techniques are so far still confined to laboratory use. More research is needed for them to become usable on a routine, daily basis, for large-scale applications.

Types of Transmissible Pathogens

Pathogens which can induce systemic infection in their host plants include viruses, viroids, phytoplasmas and some bacteria. These pathogens can be transmitted from infected mother plants to seedlings, or to planting materials such as scions, cuttings, bulbs, and tubers through vegetative propagation. In fact, vegetative propagation by tissue culture, which has become the dominant method of propagating numerous plant species, is considered to be the most efficient way of multiplying and disseminating planting materials infected with systemic pathogens. The properties of pathogens are frequently transmitted in planting materials during vegetative propagation.

Viruses

Viruses are submicroscopic pathogens that cause disease in all known organisms, from bacteria to human beings. They are obligate parasites, requiring the presence of living cells for their multiplication. Recently, viral diseases such as influenza, polio, and AIDS in human beings, and foot-and-mouth disease in cattle and other ruminants, has caused huge economic losses and had a major social impact. There have also been numerous records of plant virus diseases which have caused major economic losses. Plant viruses differ from other plant pathogens, not only in their size and morphology but also in their mode of infection and the way they multiply in host cells and are transmitted to other host plants.

Most plant viruses are very simple, possessing either RNA or DNA as the genome. This is encapsidated in a protein coat. The coat protein and genetic nucleic acid of a virus are synthesized separately in different sites in susceptible cells. At an appropriate time, they are assembled together to form progeny virus particles. Plant viruses differ from other plant pathogens in being unable to liberate themselves from the infected host plant into the environment.

Without the aid of vectors such as insects, mites, nematodes or fungi, they are also incapable of introducing themselves into their host plants through the epidermis or natural openings. However, once viruses have successfully infected the host plant, they can easily be transmitted in vegetatively propagated materials. Some plant viruses may also be transmitted from plant to plant by mechanical contact, while others may spread in infected seeds or pollen. Another characteristic of plant virus infection is that virus particles can spread through the plant, and can be detected in tissues or organs that are some distance away from the infection site. This ability to spread throughout the organism is known as systemic infection.

Although some viruses may induce distinct symptoms, most viruses show visible symptoms which are very similar to those caused by other viruses. Therefore, we cannot depend on visual inspection to identify the causal virus. Since plant viruses do not produce any structure outside the infected tissue, and are not liberated from the infected cell, different methods have been developed to detect their presence in infected cells.

Studies of the molecular biological characteristics of viruses have helped researchers to understand the function of genes in the viral genomes, and the strategies adopted by viruses for gene replication and expression of their gene products. Over the past two decades, a tremendous amount of information has been gathered on the sequence and function of viral genes. This information is now widely used for the quick identification of viruses by DNA-based technology.

Viroids

Viroids are the simplest and the most primitive of all the plant pathogens. Numerous plant diseases are induced by viroids, many of them common, such as potato spindle tuber disease and citrus exocortis. Viroids differ from viruses in being composed of only a naked single strand of ribonucleic acid (ssRNA), without the protection of a protein coat. The RNA of viroids is made up of small, molecules either rod like or a closed circle, which are highly base-paired.

Potato spindle tuber viroids have been studied in detail. They possess a serial arrangement of 26 double-stranded segments, interrupted by bulging loops of varying sizes. This arrangement of the RNA molecule gives them a characteristic stability and flexibility. In general, viroids are highly stable against thermal denaturation treatment. They are readily transmissible by mechanical contact or injury, but not by any vectors.

Like viruses, viroids can induce systemic infection in their host plants. Symptoms induced by different viroids can usually be distinguishable, and can thus be useful for diagnosis. However, it is better to use modern techniques for detection and identification, instead of depending on visual inspection alone. Since viroids contain only RNA without a coat protein, serological methods cannot be used to detect or identify them. However, they can be detected by DNA-based technology such as electrophoresis, DNA hybridization or polymerase chain reaction (PCR).

Phytoplasmas and Spirolasmas

Phytoplasmas used to be called mycoplasma-like organisms (MLOs). They are a special group of plant pathogens, similar to the mycoplasmas known to cause human and animal diseases. Mycoplasmas differ from bacteria in that they lack a cell wall or penicillin-binding sites. They are, therefore, resistant to penicillin, to which bacteria are sensitive. However, mycoplasmas are sensitive to tetracycline antibiotics. Mycoplasmas have a triple-layered plasma membrane. Because they lack a cell wall, mycoplasma cells have a characteristic lack of rigidity, and appear as pleomophic bodies.

Phytoplasmas, or MLOs, differ from animal mycoplasmas in that they cannot be isolated as pure cultures and grown in an artificial medium. However, because they are similar in morphology, and also in their sensitivity to whip-like tetracycline, these plant-infecting mycoplasmas were called MLOs until the 1990s, when `phytoplasma' was accepted as the official name of this group of pathogens. There is also a similar, related group of pathogens, the spiroplasmas. These also used to be considered MLOs until about 1970, when it was found that they could be cultured. They were also found to have a spiral movement in artificial medium. Like phytoplasmas and mycoplasmas, spiroplasmas are pleomorphic, lack a cell wall, and are sensitive to tetracycline.

Plant-infecting phytoplasmas and spiroplasmas are both found in the phloem tissue. Although they are known to induce systemic symptoms, they are usually present in a low concentration and are not distributed evenly throughout the plant. Syndromes induced by phytoplasmas and spiroplasmas include yellowing, stunting, phyllody (development of leaves in places where ovules should develop), changes in color to reddish or purple and a bunching of shoots known as witch's broom. Although some viruses may also induce symptoms of yellowing and stunting, the latter three syndromes are generally typical of infection by phytoplasmas.

Neither phytoplasmas not spidoplasmas are transmissible by mechanical contact. They can only be transmitted by vectors, particularly leafhoppers. A few can be transmitted by psyllids. In the past, the identification of phytoplasma and spiroplasma diseases relied on microscopic observation, and biological techniques such as grafting or use of insects to inoculate susceptible hosts. In the 1980s and 1990s, serological techniques, especially monoclonal antibody technology, were applied successfully. More recently, molecular techniques, including DNA probe hybridization and PCR, have been able to provide a quicker diagnosis.

Bacteria

Bacteria are prokaryotic microorganisms (i.e. they are cells that lack a membrane). Most of them are strictly saprophytic (i.e. they feed on dead organisms). They often help to decompose organic wastes in the environment. Around 1600 bacterial species are known to be pathogenic to plants. Apart from two species of Streptomyces, all plant-pathogenic bacteria are single rod-shaped cells. Most of the plant- pathogenic bacteria are motile (can move by themselves). They use flagellae to propel their bodies. These flagellae may be present singly, in groups, or distributed evenly over the cell surface. With the help of water, bacteria can move into host tissue through sites of injuries and through natural openings such as stomata. Most plant pathogenic bacteria induce local rather than systemic infection.

However, there are some specially evolved bacteria that can infect xylem or phloem tissue and become distributed throughout the plant, in a rather similar way to the systemic infection caused by viruses. In the early days, it was very difficult to culture plant pathogenic bacteria in artificial medium. For this reason, they were wrongly identified as mycoplasmas or MLO diseases for quite a long time. In the 1990s, it was found that most of them could be cultured in specially designed medium. They are now officially known as fastidious xylem- or phloem-limited bacteria, because they are found only in the xylem and phloem.

Bacteria belonging to these two groups give rise to diverse but usually species-specific symptoms in their host plants. Symptom inspection in the field is widely used for diagnosis. However, because of the low concentration of these bacteria in host tissue and the difficulty of culturing them, it is better to use serological and DNA-based techniques for quicker detection and identification.

Programme for Pathogen-free Planting Materials

The most effective way of controlling diseases that can be transmitted from mother plants to vegetatively propagated seedlings or planting materials is to develop a propagation system which ensures that all, or nearly all, of the plants produced are free of pathogens. The first step in developing a propagation system for disease-free materials is to get a complete picture of the pathogens that may attack the crop. To accomplish this needs several years of disease surveys, or consultation with experts.

The key pathogens i.e. those which are most widely distributed or those which are responsible for major reductions in yield and/or quality, should be identified and targeted. Their biological properties, disease ecology, modes of transmission, methods of field diagnosis and keys for identification should be fully understood in as much detail as possible. This step is crucial for developing a sound structure for a propagation programme that is effective in producing plants free from infection by key pathogens.

Some crops may be infected by numerous pathogens. In such cases, it may be very difficult to cover all of these, because of the cost. In real life situations, choices should be make to select key pathogens as the target to be eliminated from the propagation system.

Techniques for Pathogen Detection

The second step in developing a propagation programme for healthy planting materials is to develop techniques for detecting the key pathogens. These techniques should be accurate, sensitive, and efficient enough to allow detection as early as possible, so the pathogens can be eliminated from the propagation system. There are four main types of detection techniques.

— Biological inoculation;

— Microscopic observation;

— Serological detection; and

— DNA-based technology.

Of these four, biological inoculation is the most traditional, but is still a very accurate way of detecting pathogens. In this method, a susceptible plant host is deliberately infected with the pathogen. Biological inoculation requires a good system of different hosts which can be inoculated and observed for specific symptom expression. However, this technique requires space in a greenhouse and skilled labor to maintain the different hosts. It may take several days, or even several weeks, for symptom expression, which is usually too slow for commercial producers.

Nevertheless, biological inoculation reveals the live virus residing in the examined tissue. In contrast, the other three techniques may detect inactive viruses. This may give misleading results for certification. Therefore, biological inoculation is still used, in combination with the other techniques, in the certification of nuclear or mother stock plants. In this situation, there are usually relatively few plants to examine.

Like biological inoculation, microscopic observation requires experienced, well-trained staff and sophisticated facilities for processing. Again, it is best adopted to uses where the sample size is small rather than to large-scaled and routine inspections. Microscopic observation is therefore used mainly government institutes or specific diagnostic companies, rather than seedling producers.

It is the serological and DNA-based techniques, especially the former, which are currently widely accepted by the industry as the most efficient, dependable, reproducible and cost effective method of pathogen detection. There have been many improvements over the past two decades in serological techniques. For example, enzyme-linked immuno-sorbent assay (ELISA) has been modified into a semi-automated system that can process thousands of samples in one day.

There have also been tremendous improvements in DNA-based techniques in recent years. However, these are still used mainly by specialized laboratories with highly trained experts. Although DNA-based techniques are generally accepted as being more sensitive and specific for pathogen detection than the serological ones, they have some defects. These include the higher cost, the difficulty of reproducing results, and the possibility of false positives and cross-contamination between samples. These problems need to be solved before DNA-based techniques can be used as the routine basis of diagnosis.

Selection and maintenance of pathogen-free nuclear stock

The third step in the development of healthy plant propagation programme is to select plants which are pathogen-free, and true to type with regard to the horticultural characteristics of the species or cultivar.

Once selected, these plants should be protected from reinfection by pathogens. This is the time to decide which detection technique to use as the preliminary test to screen individual plants to make sure that they are free from infection by key pathogens. Because the number of plants for testing at this stage is usually large, techniques suitable for large-scale testing with acceptable sensitivity and cost effectiveness, such as ELISA, are desirable.

Once of this qualifying screening is completed, those selected individuals free from key pathogens should be collected and maintained in a vector-proof house for true to type selection and advanced pathogen testing. At this stage, since the number of plants for inspection should be reasonably small, more sensitive or accurate pathogen detection methods, even if they are more expensive in time and labor, can be used.

To be certain that the selected individual plants are free of all suspected pathogens, a combination of several techniques, including biological inoculation, microscopic observation, and serological and DNA-based methods are usually applied, either at the same time or in sequence. The cost, including the labor cost, is not usually the key concern at this stage of selection. It is accuracy which is the crucial point to be considered.

Furthermore, the selected mother stock plants should be carefully protected in order to prevent pathogens from entering their environment. Pathogen detection procedures should be conducted periodically to ensure that they remain free of pathogens. Every stock plant should be individually labeled, and kept well spaced from other plants to prevent mechanical contact. If there are insect vectors involved in the dissemination of pathogens, periodic pesticide applications should be made part of the management system. If re-infection is ever suspected, the suspect plant should be removed immediately from the area.

If no pathogen-free plants are available after extensive screening, this would indicate that the incidence of the pathogens concerned is already very high in the field. In this case, technology to eliminate pathogens from the mother stock plants should be used. Meristem tissue culture, thermo-inactivation treatment and many other treatments can be used for this purpose.

Pathogen-free Mother Plants

After you have established a dependable system for the maintenance of pathogen-free nuclear stock or mother plants, the next step is to consider how to mass propagate these stock plants into seedlings or planting materials for growers. One crucial point to be considered is the amplification factor. The reproduction rate of planting materials must be greater than the infection rate with the disease in the field. If the reproduction rate of pathogen-free planting materials is lower than the spread of disease in the field, the control effect of this strategy will not be acceptable to growers. Even more important, the reproduction method should be efficient, cost effective and acceptable to the industry. It cannot be considered only from the scientific point of view.

In the past, there were frequent cases of propagation methods being developed by researchers which were not accepted by commercial people, because of poor cost effectiveness and profitability. Tissue culture, for example, is a highly efficient method of reproducing planting materials. However, the cost, the high mutation rate and the possibility of re-contamination with pathogens may also be high. Where this is the case, a tissue culture system is unlikely to be acceptable to the industry or to growers.

Integration of Certification Programme

The experience of the Netherlands over several decades has shown that success in the use of healthy planting materials to control virus diseases relies at least partly on a successful certification system. This will help to give growers and importers confidence in the quality and pathogen-free status of the planting materials, so that they are willing to grow them. The fact is, the more pathogen-free planting materials are grown, the fewer the diseased plants which will exist in the fields, as they will be replaced by healthy ones. Consequently, outbreaks of disease will be suppressed.

If the government or any university could be responsible for inspecting the planting material propagation process, to guarantee the health of the plants and issue certificates, growers will probably be more than happy to grow these certified products, and abandon those may be infected. Most of those who use certified planting materials should be rewarded with better crops and less damage from disease. In fact, this system benefits not only growers but also propagators, who are more competitive if they produce certified planting materials.

However, most developed countries find that it is not easy to operate a certification system, not to mention developing ones. It takes good financial resources, trained manpower, good academic research and political support to carry out a certification system properly. Nevertheless, the experience of the Netherlands and of many other countries indicates that integrating the certification process with the propagation of pathogen-free planting materials will be the best guarantee of a successful disease control strategy.

11

Broadening Gene Pool of Rice Cultivars

Rice is the most important food crop and a staple food for 40% of the world population. More than 90% of rice is produced and consumed in Asia. It is grown under a wide range of agro-climatic conditions. Major advances have been made in increasing rice production worldwide as a result of large-scale adoption of modern high-yielding rice varieties and improved cultural practices.

The rapid rice production increase was mainly achieved through the application of principles of classical Mendelian genetics and conventional plant breeding methods. However, rice production must increase by 50 percent to meet the growing demand of this ever increasing population in 2030. To meet this goal, rice varieties with higher yield potential, durable resistance to diseases and insects, and tolerance to abiotic stresses are needed.

Plant breeding consists of two phases: the evolutionary phase and the evaluation phase. Recent advances in cellular and molecular biology and rice biotechnology have produced new tools to increase the efficiency of both phases.

Evolutionary Phase

Genetic variability for agronomic traits is the key component of breeding programs to broaden the gene pool of rice and other crops. In conventional plant breeding approaches mainly the genetic variability available within O. sativa germplasm has been exploited. However, genetic variability for many traits such as tolerance to stemborer, tungro, sheath blight and salt stress is limited in the cultivated germplasm. Breeders, therefore, search for genetic variability in other gene pools involving wild relatives of Oryza and apply new techniques for creation and transfer of variability through somaclonal variation and genetic engineering.

Wide Hybridization

Wide hybridization involving hybridization between rice and related wild species to broaden the gene pool of rice. The genus Oryza is comprised of 24 species representing AA, BB, CC, BBCC, CCDD, EE, FF, GG, HHJJ and HHKK genomes. Of these species, O. sativa (2n=24 AA) is cultivated worldwide whereas O. glaberrima (2n=24 AA) is grown in West Africa. The wild species are an important reservoir of useful genes. However, several barriers are encountered in transferring useful genes from wild species to cultivated rice. The barrier most commonly encountered is lack of crossability because of chromosomal and genic differences. Biotechnology tools such as embryo rescue and protoplast fusion have been employed to overcome the crossability barriers, and several interspecific hybrids have been produced.

Gene Transfer from Wild Species to Rice

Hybrids between rice and distantly related wild species are usually difficult to produce.

Table 1. Introgression of genes from wild Oryza species into cultivated rice.

Trait	Donor Oryza species		
	Wild species	Genome	Accession number
A. Transferred to Oryza sativa			
Grassy stunt resistance	O. nivara	AA	101508
Bacterial blight resistance	O. longistaminata	AA	-
	O. officinalis	CC	100896
	O. minuta	BBCC	101141
	O. latifolia	CCDD	100914
	O.australiensis	EE	100882
	O.brachyantha	FF	101232
Blast resistance	O. minuta	BBCC	101141
Brown planthopper resistance	O. officinalis	CC	100896
	O. minuta	BBCC	101141
	O. latifolia	CCDD	100914
	O. australiensis	EE	100882
Whitebacked planthopper resistance	O. officinalis	CC	100896
Cytoplasmic male sterility	O. sativa f. spontanea	AA	-
	O. perennis	AA	104823
	O.glumaepatula	AA	100969
Tungro tolerance	O. rufipogon	AA	105908
	O. rufipogon	AA	105909
B. Progenies under evaluation forintrogression			
Yellow stemborer	O. longistaminata	AA	-
Sheath blight resistance	O. minuta	BBCC	101141
Increased stem elongationability	O. rufipogon	AA	CB751
Tolerance to acidity, iron and aluminum toxicity	O. glaberrima	AA	many
Weed competitiveness	O. glaberrima	AA	many

Low crossability and abortion of hybrid embryos are the common features of such crosses. Hybrids have been produced through embryo rescue between elite breeding lines or varieties of rice and several accessions of wild species representing BBCC, CC, CCDD, EE, FF, GG, HHJJ, and HHKK genomes. A number of useful genes for resistance to brown planthopper (BPH), whitebacked planthopper (WBPH), bacterial blight (BB), blast and tungro disease have been transferred from wild species to rice (Table 1).

The first example of transfer of a useful gene from wild species is the introgression of a gene for grassy stunt virus resistance from O. nivara to cultivated rice. Grassy stunt resistant varieties, IR28, IR29 and IR30 were released for cultivation in 1974 and IR32, IR34, IR36 were released in later years. Grassy stunt resistance gene has now been incorporated into numerous varieties.

Cytoplasmic Diversification

The wild species with AA genome have been an important source of cytoplasmic male sterility (CMS), the major tool for breeding commercial rice hybrids. Lin and Yuan reported the development of male sterile line having cytoplasm of wild species (O. Sativa l. F. Spontanea) and nuclear genome of rice. This cytoplasmic source has been designated as wild abortive (WA) refering to a male sterile wild rice plant having abortive pollen. About 95% of the male sterile lines used in commercial production in China and other countries have WA type of cytoplasm.

A new CMS source from O. Perennis was transferred into indica rice. Newly developed CMS line has been designated as IR66707A. It has the cytoplasm of O. Perennis and the nuclear genome of cultivated rice IR64.

Genetic studies show that male sterility source of IR66707A is different from that of WA. Another CMS line (IR 69700A) having cytoplasm of O. Glumaepatula and nuclear genome of IR64 has been developed. Both IR66707A and IR69700A are completely stable for male sterility. Search for restorers of these CMS sources is underway.

Tagging Alien Genes

Alien genes for resistance to BPH, BB, and blast have been tagged with molecular markers. RFLP analysis of the introgression lines derived from O. sativa x O. officinalis showed introgression of the chromosome segments in 11 of the 12 chromosomes of O. officinalis. Most introgressed segments were detected by single RFLP markers and the flanking markers were negative for introgression. RFLP analysis was carried out on the introgression line IR65482-4-136-2-2 with resistance to three BPH biotypes derived from O. sativa x O. australiensis.

Of the 14 probes, previously mapped to chromosome 12, only RG457 detected introgression from O. australiensis. Co-segregation for BPH resistance and RG457 showed that the gene for resistance to BPH is linked to molecular marker RG457, with a crossover value of 3.68 + 1.29%. Such tight linkage should facilitate selection for BPH resistance during the transfer of this resistance to other elite breeding lines of rice.

Somaclonal Variation

Somaclonal variation refers to the variation arising through tissue culture in regenerated plants and their progenies. The technique consists of growing callus or cell suspension cultures for several cycles and regenerating plants from such long term cultures. The regenerated plants and their progenies are evaluated in order to identify individuals with a new phenotype. Some useful somaclonal variants, including those for disease resistance and male sterility, have been isolated.

Genetic Engineering

Introduction of alien genes into crop species allows plant breeders to accomplish breeding objectives considered impossible until a decade ago. Transformation of rice is now possible through several techniques, e.g. electroporation, polyethylene glycol induced uptake of DNA into protoplasts, and microprojectile bombardment. More recently Agrobacterium mediated transformation has become feasible. Transgenic rices carrying agronomically important genes for resistance to stemborer, fungal pathogens, and herbicide tolerance have been produced in japonica as well as in indica rices (Table 2).

Table 2. Some examples of transgenic rice plants with agronomically important genes.

Transgene	Transfer method	Useful trait
bar	Microprojectile bombardment	Tolerance to herbicide
bar	PEG- mediated	Tolerance to herbicide
Coat protein	Electroporation	Tolerance to stripe virus
Coat protein	Particle bombardment	Tolerance to rice tungro spherical virus
Chitinase	PEG-mediated	Sheath blight resistance
cryIA(b)	Electroporation	Tolerance to striped stemborer
cryIA(b)	Particle bombardment	Tolerance to yellow stemborer and striped stemborer

contd...

contd...

cryIA(c)	Particle bombardment	Tolerance to yellow stemborer
cry1A(b), cryIA(c)	Agrobacterium mediated	Tolerance to striped stemborer and yellow stemborer
cry1A(b), cryIA(c)	Particle bombardment	Tolerance to yellow stemborer
cry1A(c), cry2A, gna	Particle bombardment	Tolerance to stemborer, leaf folder and brown planthopper
CpTi	PEG-mediated	Tolerance to striped stemborer and pink stemborer
gna	Particle bombardment	Insecticidal activity for brown planthopper
Corn cystatin (CC)	Electroporation	Insecticidal activity for Sitophilus zeamais
Xa21	Particle bombardment	Resistance to bacterial blight
codA	Electroporation	Increased tolerance to salt
Soybean ferritin	Agrobacterium mediated	Increased iron content in seed
psy	Particle bombardment	Phytoene accumulation in rice endosperm
psy, crt1, lcy	Agrobacterium mediated	Provitamin A

Several laboratories have produced transgenic rices mainly through protoplast mediated DNA transformation but also via microprojectile bombardment. Hiei et al. and Cheng et al. obtained a large number of fertile transgenic plants through Agrobacterium mediated transformation. Chen et al. produced transgenic rice carrying multiple transgenes. Transgenic plants were produced after cobombarding embryogenic callus and cell suspensions with a mixture of 14 different pUC-based plasmids. Eighty-five percent of the Ro plants contained more than 2 and 17% more than nine of the target genes. Plants containing multiple transgenes had normal morphology and 63% set viable seeds.

Transgenic Rice for Modifying Yield

ADP-glucose pyrophosphorylase (ADPGPP) is critical enzyme in regulating starch biosynthesis in plant tissues. Starch levels and dry matter accumulation were enhanced in potato tubers of plants transformed with glgC[16] gene from E. coli encoding ADPGPP. Yield potential of these lines will be evaluated shortly. Also, the transfer of C_4 traits into C_3 rice is being explored to improve photosynthetic efficiency. Ku et al. used Agrobacterium-mediated transformation and introduced into rice a gene for phosphoenolpyruvate carboxylase (PEPC) from maize, which catalyzes the initial fixation of atmospheric CO_2 in C_4 plants.

Most transgenic rice plants showed high-level expression of the maize gene; the activities of PEPC in leaves of some transgenic plants were two-to threefold higher than those in maize, and the enzyme accounted for up to 12% of the total leaf soluble protein. Level of expression of the maize PEPC in transgenic rice plants correlated with the amount of transcript and the copy number of the inserted maize gene. The transgenic rice plants exhibited reduced O_2 inhibition of photosynthesis and photosynthetic rates comparable to those of untransformed plants. These findings demonstrate a successful strategy for introducing the key biochemical component of the C_4 pathway of photosynthesis into rice.

Transgenic Rice for Insect Resistance

As early as 1987, genes coding for toxins from Bacillus

thuringiensis (Bt) were transferred to tomato, tobacco and potato to provide protection against lepidopteran insects. A major target for Bt deployment in transgenic rice is the yellow stemborer, Scirpophaga incertulas (Walker). The pest is widespread in Asia. Fujimoto et al. introduced a truncated -endotoxin gene, *cry1A(*b) into rice. Transgenic plants in the R_2 generation expressing the *cry1A*(b) protein had increased resistance to striped stemborer and leaffolder. Wunn et al. introduced *cry1A*(b) gene into IR58 through particle bombardment.

The transgenic plants in R_0, R_1 and R_2 generation showed significant insecticidal effect on several lepidopteran insect pests. Feeding studies showed up to 100% mortality for yellow stemborer and striped stemborer larvae. Nayak et al. transformed IR64 through particle bombardment using *cry1A*(c) gene. Six independent transgenic lines showing high expression of insecticidal crystal protein were identified. The transferred synthetic *cry1A*(c) gene was stably expressed in the T2 of these lines and the transgenic rice plants were highly toxic to yellow stemborer larvae. Cheng et al. produced over 2600 transgenic rice plants in nine strains through Agrobacterium-mediated transformation. The plants were transformed with two synthetic *cry1A*(b) and *cry1A*(c) coding sequences from Bacillus thuringiensis. Bioassays in R_1 indicated that the transgenic plants were highly toxic to striped stemborer and yellow stemborer. Maqbool et al. transformed, Basmati 370 and M7 cultivars using the novel endotoxin *cry2A* gene.

Disease Resistance

Sources of resistance to some diseases have been identified within cultivated rice germplasm. However, sources of resistance to sheath blight are not available and only a few donors for resistance to tungro disease. A highly successful strategy, termed coat protein (CP) mediated protection, has been employed against certain viral diseases such as tobacco mosaic virus in tobacco and tomato. A coat protein gene for rice stripe virus was introduced into two japonica varieties by electroporation of the protoplasts. The resultant transgenic plants expressed the CP at high levels (up to 0.5% of the total soluble proteins) and exhibited a significant level of resistance to virus infection. The resistance was inherited to the progenies.

Zhang et al. introduced *Xa21* using projectile bombardment of cell suspensions of elite indica rice varieties; IR64, IR72, Minghui 63, and BG90-2. Six of the 55 R_0 lines carrying *Xa21* showed high level of resistance to BB in subsequent generations. Sheath blight of rice is caused by the fungus Rhizoctonia solani, which has a wide host range. About six chitinase genes have been identified in rice and are being manipulated to increase the level of resistance to fungal diseases. Lin et al. introduced 1.1 kb rice genomic DNA fragment containing a chitinase gene through PEG mediated transformation.

Abiotic Stress Tolerance

Genetic engineering approaches hold great promise in developing rice cultivars with higher levels of tolerance to abiotic stresses. Sakamoto and Murata introduced *codA* gene for choline oxidase from Arthrobacter globiformis into rice. The codA gene was inherited to the second generation of transgenic rice and its expression was stably maintained at levels of the mRNA, the protein and enzyme activity. Levels of glycinebetaine were estimated as high as 1 and 5 μmol g^{-1} fresh leaves in two types of transgenic plants (chl COD and cyt COD plants) in which choline oxidase was targeted to the chloroplasts and cytosol respectively. Preliminary results showed that the transgenic plants grew better than the control plants during the recovery from treatment with 0.15 M NaCl for 7 days. Further analysis of transgenic plants showed their ability to synthesize betaine conferring enhanced tolerance to salt and cold.

Improved Nutritional Quality

Millions of rice consumers who depend on rice for a large proportion of their calories suffer from vitamin A deficiency. Ye et al. produced transgenic rice (golden rice) with the provitamin A (β-carotene) biosynthetic pathway engineered into its endosperm. Agrobacterium mediated transformation was used to introduce three genes; phytoene synthase (psy), phytoene desaturase (crt1) and lycopene cyclase (lcy). HPLC analysis showed presence of β-carotene in transgenic seeds. The transformed rice Taipei 309 is no longer cultivated. Efforts are now underway at IRRI to transfer the genes for β-carotene into IR64 and others.

Goto et al. introduced entire coding sequence of the soybean ferritin gene into kita-ake, a rice cultivar, via Agrobacterium mediated transformation. The introduced ferritin gene was regulated by the rice seed storage protein glutelin promoter, GluB-1 and terminated by the Nos polyadenylation signal. Synthesis of soybean ferritin protein was confirmed in each of the transformed rice seeds by western blot analysis. Specific accumulation in endosperm was determined by immunological tissue printing. The iron content of T1 seeds was upto three times more than the untransformed seeds.

Evaluation Phase

Anther Culture

Anther culture is important technique to develop true

breeding lines in the immediate generation, thereby shortening the breeding cycle of new varieties. The selection efficiency in doubled haploid (DH) lines is higher, especially when dominance variation is significant. Seeds of such populations can be distributed to workers in other laboratories, and populations can be grown repeatedly in many different environments, greatly facilitating the additional mapping of both DNA markers and genetic loci controlling traits of agronomic importance. In rice, mapping populations (DH lines) produced through anther culture of IR64 x Azucena (indica/japonica cross), and CT9993 x IR62266 are being used in molecular mapping of genes including quantitative trait loci.

A number of varieties and improved breeding lines have been developed through anther culture in China, Republic of Korea, Japan and U.S.A. Most of the anther culture derived varieties are japonicas. Indica rices are generally regarded as recalcitrant for anther culture. At IRRI, anther culture technique is being employed to obtain DH lines for different rice breeding objectives. One of the anther culture derived lines, IR51500-AC-11-1, has been named as a variety (PSBRc50) in Philippines. Many DH lines produced through anther culture are now being used as parents in breeding programs. DH lines are being produced to overcome sterility in the progenies of O. sativa x O. glaberrima crosses.

Molecular Marker Technology

A series of molecular markers, e.g. random fragment length polymorphism (RFLP), random amplified polymorphic DNA (RAPD), amplified fragment length polymorphism (AFLP) and microsatellites have become available. These markers offer new opportunities for various studies in genetic and breeding research particularly in the construction of saturated molecular maps, gene tagging, QTL mapping, marker-assisted breeding, gene pyramiding, physical mapping of genome, map-based gene cloning, alien introgression and DNA finger printing of pathogen populations.

Molecular Maps

The availability of comprehensive molecular genetic map in rice comprising of more than 2300 DNA markers has been a major advance in rice genetics. A molecular genetic map of rice, based on RFLPs, was developed at Cornell University, Ithaca, USA, in collaboration with IRRI. This map was based on F2 population derived from an indica x japonica cross and mapped sequences were cloned from a genomic library derived from the indica variety IR36. Later many new markers were added to the map. A comprehensive molecular genetic map consisting of 2275 DNA markers has been developed at the Rice Genome Research Program in Japan. Singh et al. mapped centromeres on the molecular genetic map of rice and determined correct orientation of linkage groups.

Molecular Marker Aided Selection (MAS)

Most genes of economic importance behave in a dominant or recessive manner and require time consuming efforts to transfer. Sometimes the screening procedures are cumbersome and expensive and require large field space. If such genes can be tagged by tight linkage with DNA or isozyme markers, time and money can be saved in transferring these genes from one varietal background to another. A molecular marker very closely linked to the target gene can act as a "tag" which can be used for indirect selection of the gene(s) in breeding program.

Gene Tagging in Rice

Development of varieties with durable resistance to rice blast caused by Pyricularia oryzae, and bacterial blight caused by Xanthomonas oryzae pv. Oryzae is the focus of a co-ordinated effort at IRRI and National Agricultural Research Systems (NARS) using molecular marker technology. Several genes for bacterial blight and for blast resistance have been tagged with molecular markers (Table 3).

Table 3. Some examples of rice genes of agronomic importance mapped with molecular markers

Gene	Trait	Chromosome	Link marker
Pi1	Blast resistance	11	Npb181
Pi2	Blast resistance	6	RG64
Pi4	Blast resistance	12	RG869
Pita	Blast resistance	12	RZ397
Pi5	Blast resistance	4	RG498 , RG788
Pi6	Blast resistance	12	RG869
Pi7	Blast resistance	11	RG103
Pi9	Blast resistance	6	RG16
Pi10	Blast resistance	5	RRF6, RRH18
Pi11	Blast resistance	8	BP127
Pi12	Blast resistance	12	RZ869
Pib	Blast resistance	2	RZ123
Pi20	Blast resistance	12	XNbp88
Pikm	Blast resistance	11	R1506
Pb1	Blast resistance	11	S723
Xa1	Bacterial blight resistance	4	Npb235
Xa2	Bacterial blight resistance	4	Npb235 Npb197
Xa3	Bacterial blight resistance	11	Npb181 Npb78
Xa4	Bacterial blight resistance	11	Npb181
xa5	Bacterial blight resistance	5	RG556
Xa10	Bacterial blight resistance	11	OP07
xa13	Bacterial blight resistance	8	RZ390 RG136
Xa21	Bacterial blight resistance	11	RG103

contd...

contd...

Xa22	Bacterial blight resistance	1	RZ536
Stvb1	Rice stripe virus resistance	12	Npb220
RTSV	Rice tungro spherical virus resistance	4	RZ262
RYMV	Rice yellow mottle virus resistance	12	RG341
Bph1	Brown planthopper resistance	12	XNpb 248
bph2	Brown planthopper resistance	12	G402
Bph9	Brown planthopper resistance	12	S11679
Bph10	Brown planthopper resistance	12	RG457
Bph	Brown planthopper resistance	-	MRF0P16938
Qbph1	Brown planthopper resistance	3	R1925
Qbph2	Brown planthopper resistance	4	C820
ef	Early flowering	10	CDO98
fgr	Fragrance	8	RG28, RM223
Wph1	Whitebacked planthopper resistance	7	
WBPH	Whitebacked planthopper resistance	11	RG103
Gm1	Gall midge resistance	-	OPK7
Gm2	Gall midge resistance	4	RG329, RG214
gm3	Gall midge resistance	-	OPQ12
Gm4	Gall midge resistance	8	OPM12
Gm5	Gall midge resistance	12	OPB14
Gm6	Gall midge resistance	4	RG214
Rf1	Fertility restoration	10	OSRRF
Rf2	Fertility restoration	1	CDO686/RZ58
Rf3	Fertility restoration	1	RG532
Rf5	Fertility restoration	1	RG374
S5	Wide compatibility	6	RG213
Se1	Photoperiod sensitivity	6	RG64
Se3	Photoperiod sensitivity	6	A19
sdg	Semidwarf	5	RZ182
sd1	Semidwarf	1	RG109
tms1	Thermosensitive male sterility	8	-
tms3	Thermosensitive male sterility	6	OPAC3640
tms4	Thermosensitive male sterility	2	RM27
rtms1	Thermosensitive reverse male sterility	10	RM239

contd...

contd...

pms1	Photoperiod sensitive male sterility	7	RG477
pms2	Photoperiod sensitive male sterility	3	RG191
pms3	Photoperiod sensitive male sterility	12	C751/RZ261
Sub1	Submergence tolerance	9	RZ698
Salt	Salt tolerance	7	RG64
OSA3	Salt tolerance	12	RG457

Resistance, based on single major genes, has generally been overcome in a few years in blast prone environments. Resistance in some cultivars, however, has been more long lasting or durable. In several cases, durable resistance to blast is believed to be associated with quantitative or polygenic inheritance. Under these conditions, there is little or no gain in fitness for a pathogen variant to overcome only a fraction of the polygenes. Breeders now aim at incorporating quantitative or polygenic resistance into rice varieties.

Wang et al. identified a gene Pi5 conferring complete resistance including nine regions of the genome with quantitative effects on blast resistance. These latter were considered to be putative QTL for blast resistance. Analyses of major resistance genes has also been carried out using near isogenic lines (NILs). The NILs have been used for mapping of the genes Pi2 and Pi4.

Efforts to detect markers closely linked to bacterial blight resistance genes have similarly taken advantage of the availability of a set of ten NILs. These lines each contain a single gene for resistance and, in the case of *Xa21*, the gene has been introgressed from the wild species of rice, O. longistaminata. Segregating populations have been used to confirm co-segregation between the RFLP markers and *Xa1, Xa2, Xa3, Xa4, xa5, Xa10, xa13 and Xa21.*

Genes for aroma, wide compatibility, fertility restoration and thermosensitive genetic male sterility, BPH resistance, tolerance to tungro virus and submergence have been tagged with molecular markers (Table 3). PCR based markers have been developed for some of these traits to facilitate selection. Nair et al. developed PCR-based markers for gall midge resistance genes *Gm2* and *Gm4* which are useful in MAS for developing gall midge resistant varieties. Ichikawa et al. developed PCR based marker for Rf1, a fertility restorer gene. The emphasis now is on tagging genes of economic importance by developing PCR based markers.

QTL Mapping

Genetic differences affecting such as yield, quality and tolerance to various abiotic stresses (drought, salinity, submergence, etc.) are controlled by a relatively large number of loci, each of which can make a small positive or negative contribution to the final phenotypic value of the traits. Such loci are termed "quantitative trait loci or QTL". The genes governing such traits called polygenes or minor genes also follow Mendelian inheritance but are greatly influenced by the environments.

The advent of molecular markers has made it possible to map the QTL having large genotypic effect on phenotype. These molecular markers allow transformation of QTL into Mendelian or quasi-Mendelian entities that can be manipulated in classical breeding programmes. Through linkage analysis of large segregating populations polymorphic for molecular markers and QTL for yield and yield components, locations of the QTL can be determined. Several QTL for traits such as blast resistance, root traits, submergence tolerance and yield components, have been mapped with molecular markers.

Wang et al. mapped *Pi5* and *Pi7* for blast resistance on chromosomes 4 and 11 respectively. Nine QTL having quantitative resistance to isolate PO6-6 of blast were also identified. Li et al. identified 6 QTL contributing to resistance to Rhizoctonia solani. These QTL are located on 6 of the 12 rice chromosomes and accounted for 60% of the genotypic variation in Lemont x Tequing cross. Nandi et al. used AFLP markers and identified 4 QTL for submergence tolerance on chromosomes 6, 7, 11 and 12 of rice.

In addition, a major gene *Sub1* for submergence tolerance was localized on chromosome 9. Xiao et al. analyzed BC2 test cross families from the interspecific cross (O. sativa x O. rufipogon) and found that O. rufipogon alleles at marker loci RM5 on chromosome 1 and RG256 on chromosome 2 were associated with an 18 and 7% increase in grain yield per plant. A total of 68 significant QTL were identified, and of these 35 (51%) had beneficial alleles derived from the phenotypically inferior O. rufipogon parent. Moncada et al. used advanced backcross QTL analysis on O. sativa x O. rufipogon derivatives and found that certain regions of rice genome harbor genes useful over range of environments.

Molecular markers can be used to identify QTL from wild species responsible for transgressive segregation. Comparative mapping among different cereals has also increased the efficiency of mapping orthologous QTL. Future research should focus on (1) identification of QTL which could be exploited in different environments, (2) exploitation of complementary QTL to isolate transgressive segregants particularly from interspecific crosses, (3) identification of orthologous QTL among different species, as the conservation of such QTL among species may provide new opportunities for manipulation of economic traits, (4) high resolution of QTL to determine whether QTL are single genes or clusters of tightly linked genes and whether overdominance plays a significant role in heterosis, (5) cloning of QTL based on high resolution mapping.

Gene Pyramid

The development of saturated molecular maps and the possibility of finding tight linkage of target genes with molecular markers such as RFLPs and conversion of these markers to PCR-based markers have provided new opportunities to use MAS in rice breeding. Protocols for PCR based MAS in rice have been developed. In MAS individuals carrying target genes are selected in a segregating population based on tightly linked markers rather than on their phenotypes. Thus, the populations can be screened at early seedling stage and in various environments.

MAS helps overcome interference from interactions between different alleles of a locus or of different loci. MAS was applied for pyramiding four genes for BB resistance e.g. *Xa4, xa5, xa13,* and *Xa21*. Breeding lines with two or three genes were also developed. The pyramided lines showed a wider spectrum and a higher level of resistance than lines with only a single gene. Yoshimura et al. developed lines carrying *Xa4* + *xa5* and *Xa4* + *Xa10* using RFLP and RAPD markers linked to the BB resistant genes. These lines were evaluated for reaction to 8 strains of BB, representing eight pathotypes and three genetic lineages. Lines carrying *Xa4* + *xa5* were more resistant to isolates of race 4 than were either of the parental lines.

Sanchez et al. used sequence-tagged site (STS) markers to pyramid three genes for BB resistance in an elite breeding line of rice. The pyramided lines having three or four genes in combination showed an increased and wider spectrum of resistance to bacterial blight than those having a single resistance gene. Singh et al. also used MAS to pyramid genes for BB resistance into a high-yielding indica rice cultivar, PR106, that is susceptible to BB. MAS was also employed to pyramid genes for resistance to blast and gall midge. Molecular marker analysis is used to identify pathogen populations with wide diversity, which are then employed for screening and for developing durable resistant genotypes. Gene deployment strategies are being employed based on understanding of pathogen population genetics and on the genetic basis of durable resistance.

Synteny Relationships

The development of molecular genetic maps has helped in understanding the homoeologous relationships between the genomes of crop plants. Kurata et al. analyzed synteny relationships between the genomes of rice and wheat and found that many wheat chromosomes contained homoeologous genes and genomic DNA fragments in the same order as in rice. Comparative genome mapping in rice, maize, wheat, barley, sorghum, foxtail millet, and sugarcane demonstrates that gene content and order are highly conserved at both the map and megabase levels between different species within the grass family, but the amount and organization of repetitive sequences have diverged considerably.

Based on comparative mapping information, gene location in one species can be used to predict the presence and location of orthologous loci in other species. The synteny relationships among cereals have resulted in the discovery of common genes such as dwarf phenotype in maize (*D8*) and wheat (*Rht1*) based on genomic information derived from rice.

Map based Cloning of Genes

The high density molecular genetic map coupled with the development of BAC and YAC libraries has led to the isolation of rice genes such as *Xa1, Xa21,* and *Pib.* Wang et al. used BAC library and identified clones linked to *Xa21* gene for BB resistance. Song et al. isolated Xa21 gene by positional cloning. The isolated gene has been introduced into rice cultivars through transformation. Similarly, Xa1 for BB resistance has been cloned through map-based cloning strategy. The expression of Xa1, unlike previously studied resistance genes, was induced by wound and pathogen infection.

At IRRI, BAC library has been constructed from IR64 genomic DNA. The library consists of 18,432 clones. BAC clones carrying *xa5* for BB resistance have been identified. Efforts to clone xa5 are underway. Once cloned, this gene would be transferred to elite rice varieties through transformation. Wang et al. cloned blast resistance gene Pib. The expression of Pib is induced upon altered environmental conditions such as altered temperature and darkness. Sanchez et al. identified BAC contigs flanking the xa13 locus for BB resistance.

Functional Genomics

Rice has become a model system for genomics research due to the comparatively its smaller size of genome, the synteny of its genome with those of other cereals, the availability of densely populated molecular maps containing more than 2,300 DNA markers, characterized YAC and BAC libraries, large-scale analysis of expressed sequence tags (ESTs), the vast amount of genetic resources, and the comparative ease of transformation. Mutants induced through transposable elements, retrotransposons, miniature inverted repeat transposable elements (MITEs), and T-DNA insertions have provided a wealth of genetic resources for functional genomics.

Some notable examples are T-DNA-tagged insertional mutants with 30,000 lines carrying 42,000 T-DNA inserts, Tos17 retrotransposon insertional mutants with about 30,000 lines carrying more than 250,000 Tos17s, and deletion mutant populations produced by fast-neutron, gamma radiation, and chemical mutagenesis. These plant materials have provided the necessary link between structural or sequence data and gene function through forward and reverse genetics. Furthermore, a large number of ESTs from various cDNA libraries have been produced for microarray expression experiments. One of the major challenges is to determine the function of previously unknown rice genes revealed by sequencing projects.

New high-throughput methods are being developed for expression analysis. Biochips are being used to follow changes in gene expression in response to abiotic stresses. Using gene chips or microarrays, the representative genes of rice can be analyzed on a glass slide and used in RNA hybridization to reveal gene expression patterns and identify pathways by association. A combination of mRNA and proteomics will precisely reveal the function of rice genes involved in the phenotypic expression of different agronomic traits.

Efficiency of Evalutionary Phases of Rice Breeding

Recent advances in cellular and molecular biology of rice offer new opportunities to enhance the efficiency of both evolutionary and evaluationary phases of rice breeding.

Biotechnology is becoming important component of rice breeding (Tables 4, 5). Anther culture has been used by breeders to shorten the breeding cycle for developing rice varieties, fix recombinants by overcoming sterility in distant crosses. Several rice varieties have been developed through anther culture. Doubled haploid populations are important in mapping of genes governing agronomic traits including QTL.

Table 4. Overcoming some of the constraints of conventional rice breeding using biotechnology tools

Constraint	Biotechnology applications
Limited genetic variability for resistance to stemborers	Transgenic rice carrying Bt genes show enhanced resistance to stemborers
Limited genetic variability for resistance to sheath blight	Transgenic rice carrying chitinase genes show tolerance to sheath blight
Lack of genetic variability for β-carotene	Transgenic 'golden' rice carrying phytoene synthase (psy), phytoene desaturase (crt1) and lycopene cyclase (lcy) show pro-vitamin activity in rice seeds
Low selection efficiency to pyramid genes for durable resistance to pests	Marker assisted selection practised to pyramid genes for bacterial blight, blast and gall midge resistance
Longer breeding cycle of rice varieties	Varieties developed through anther culture in shorter period
Narrow gene pool for resistance to pests	Genes from wild species with broad spectrum of resistance to bacterial blight, brown planthopper and tungro incorporated into elite breeding lines of rice
Characterization of pathogen population-difficult and laborious	DNA fingerprinting practised to characterize pathogen populations and for gene deployment for durable resistance

Table 5. Some examples of the biotechnology products in rice

Biotechnology tools	*Product(s)*
Anther culture	Several improved cultivars and elite breeding lines developed and released for commercial cultivation in China, Korea and Philippines.
Molecular marker assisted selection	Marker assisted selection practiced.Pyramided lines with durable resistance carrying genes for resistance to bacterial blight, blast and gall midge developed in India, China, Philippines and Indonesia. Some of the pyramided lines have been field tested and are in the process of release for commercial cultivation.
Alien introgression	Elite breeding lines of rice carrying genes from wild species for resistance to bacterial blight, blast, brown planthopper, tungro and tolerance to acidic conditions have been developed. Varieties resistant to brown planthopper and tolerant to acidity have been released in Vietnam.
Somaclonal variation	Early maturing, high yielding and blast resistant varieties released in Hungary and Japan.
Transformation	Novel genes inserted into rice cultivars for tolerance to herbicide and resistance to stemborer, sheath blight and bacterial blight and pro-vitamin A for improved nutritional quality. Field tests of transgenic rice carrying Bt gene for stemborer resistance have been made in China. However, no commercial release of transgenic rice has been made so far.

Molecular markers have led to the tagging of numerous genes for tolerance to major biotic and abiotic stresses. MAS has become a tool for moving genes from one varietal background to other and for pyramiding genes and to develop durable pest resistant cultivars. Efforts on fine mapping of QTL should provide means to pyramid QTLs for tolerance to major abiotic stresses. Map based cloning has made it possible to isolate useful genes governing important agronomic traits and incorporation of these genes into elite rice cultivars through transformation.

Advances in tissue culture and molecular marker technology have resulted in broadening of the gene pool of rice and enhanced the efficiency of introgression of useful genes from wild species across crossability barriers. Advances in genetic engineering have facilitated the introduction of novel genes into rice through transformation. Transgenic rices with enhanced resistance to diseases and insects and improved nutritional quality have been produced and will impact rice production and in providing more nutritious rice. Identification of genes and their manipulation would be another major breakthrough in rice genetics and breeding.

Future food and nutritional security will depend upon availability of rice cultivars with higher yield potential, durable resistance to diseases and insects, tolerance to abiotic stresses and higher levels of micronutrients in the grain. Conventional breeding methods and biotechnology tools will help us meet these challenges.

12

Crossbreeding

Crossbreeding is the process of creating hybrids. Hybrids can be formed by cross breeding two different breeds, races, variants, or species. The nature of the hybrid can be used to determine the relationship of the two parents: if the hybrid is able to reproduce, then the parents were both members of the same species; if the hybrid is sterile (unable to reproduce), then the parents were members of different species.

The first type of hybrid, the crossbreed, is very common. If you cross a broccoli (*Brassica oleracea*) with a cauliflower (*Brassica oleracea*) you get a "broccoflower" which can be planted, grown, and harvested just like its parents. Similarly, if you cross a Cocker Spaniel (*Canis familiaris*) with a Poodle (*Canis familiaris*), you get a "Cockerpoo", one of the more common mutt breeds in america. Since both parents belong to the same species, the offspring belong to that species, i.e. all dogs are very closely related, no matter how different they look.

The second type of hybrid, the true hybrid, is much less common, because it is unusual for two different species to be closely enough related. The example you gave is a good one: a mule is the offspring of a male donkey (*Equus asinus*) and a female horse (*Equus caballus*);

all mules are sterile, so the only way to make more mules is to cross more horses and donkeys. If you cross a female donkey and a male horse, you get a hinny, which is also sterile.

There are also some hybrid plants, like hybrid roses, which can only be reproduced from clippings of the original plant and don't produce viable seeds. In both cases, the two organisms crossed must be very closely related to produce any offspring. Generally this means that they must at least be in the same genus, although often that is not even enough. In the same way, crossing animals from different genera will not produce offspring, and certainly crossbreeding an animal and a plant is impossible.

There are very many situations in which the ideal producing animal is some intermediate between a tropical adapted and an improved temperate breed. This applies particularly to dairy cattle kept under intensive conditions in the humid tropics and in many parts of the wet-dry tropics and to beef cattle under range conditions in the wet-dry tropics and the subtropics.

Crossbred Population

When it has been established that in a given environment (including climate, disease situation, feeding and management systems) the most productive animal is one containing a proportion of local blood and a proportion of imported genes, then an appropriate breeding system must be chosen which will maintain this intermediate type after the initial crossing. There are several possibilities.

New Breed Formation

This is also called the formation of a gene pool. After the

first cross, F_1 animals are mated together to form an F_2 followed by an F_3, and so on. If the desired proportion of European blood is 75 percent then a backcross is made before the inter se mating starts. The advantage of the gene-pool approach is that the population is self-replacing and after the initial cross no further outside blood is needed.

However, because of the former emphasis in Europe and America on the importance of purebreeding and on uniformity within a breed, there has in the past been reluctance to take the step of attempting to form a new breed especially in a species with a long generation interval. People are afraid that breeding from crossbreds will lead to excessive variation, and therefore that formation of a new breed is only possible if immense resources of animals are available. In fact, the increase in variation on breeding an F_2 generation has been much exaggerated.

To begin with, it is only the genetic variation which is increased, and with characters of low heritability this is only a small part of the total variation. Secondly, the obvious effects are due to segregation of colour, colour pattern and morphological characters (e.g. horns) which are controlled by single genes. The increase in variation between F_1 and F_2 for such quantitative characters as growth rate and milk yield will be very small and, in view of other sources of variation, may not be detectable.

Certainly there is a danger that if the initial crossing is not done on a large enough scale then problems of inbreeding may be encountered. This is particularly the case if the breed is based on a single herd or on only one or two imported sires. Inbreeding leads to decline in fertility, vaibility and growth rate to a greater or lesser extent. This has often not been realised and inbreeding has been deliberately employed in order to concentrate the blood of outstanding sires or to produce uniformity in a new breed. But uniformity is an elusive goal, a will-o'-the-wisp. It must not be sought at the expense of productivity. Indeed it is essential at the outset to have as much variation as possible in order to allow scope for selection.

The aim of crossbreeding is to combine the high yield of the European breed with the resistance of the local breed, and intense selection is needed to find the few animals combining both characters. If a breed is to be formed in a single experimental herd it is therefore essential to use a large number of unrelated sires (at least 10-15) in the formative stages. As early as possible the programme should be extended to other herds and sires should be exchanged between herds. In the case of dairy cattle a large cow population will be needed for progeny testing the bulls. Above all, animals should not be selected according to colour, conformation or other fancy points but for important economic characters.

Where resources for intense selection are not available it would be easier to import a new intermediate breed from elsewhere and grade the local population to this imported breed. Unfortunately such new breeds are rare. The advantage of the single breed approach is that in an unsophisticated social environment it is easier for the farmer to deal with one type of animal. He can either keep his own sire or obtain semen from an AI service if available. There is no question of different sires for different females which is the essence of a crossbreeding system.

The major disadvantage of new breed formation is that it is a long-term programme and by the time it yields results the environment or market demand may have changed so that a different type of animal is needed. A systematic crossbreeding system is much more flexible.

Systematic Crossbreeding System

The essence of a systematic crossing system is that breeding animals of both pure breeds involved (local adapted and temperate improved) are used in each generation. There are two main categories - terminal crossing and rotational crossing.

Terminal Crossing

In this type of cross the progeny are not used for breeding. It is therefore not suitable for dairy animals but is important for meat animals. There are several varieties:

a. Re-creation of the F_1 in each generation. Since it is the commercial F_1 animals which are produced every generation this is also called "commercial crossing". Both sexes are slaughtered (for meat) without being used for breeding.

b. Three-way crossing

Commercial crossing exploits only the hybrid vigour in the growing animal. In order to cash in also on that in maternal ability a second stage of crossing is needed so as to breed from a crossbred dam. This is very important because a major proportion of the total advantage may come from the use of crossbred dams. The second cross may be a backcross to the sire breed or a cross to a third breed.

This system has its maximum advantage if the crossing is stratified according to the environment. In Britain, for instance, local breeds of sheep are maintained on the mountains; they are crossed with a hardy improver breed and the crossbreds are kept in the hills; the crossbred ewes are crossed with a second improver breed and the second-cross lambs fattened in the lowlands. A similar scheme could be applicable in the tropics with the environment being superior in each phase in terms of rainfall, temperature or feeding level rather than altitude.

Rotational Crossing

In crisscrossing (or reciprocal backcrossing) sires of the temperate and tropical breeds are used in alternate generations (Figure 1).

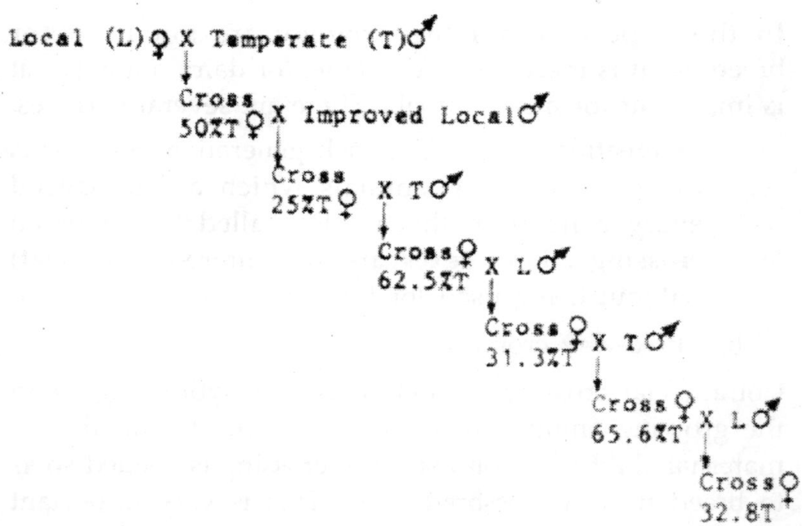

Figure 1. Crisscrossing (reciprocal backcrossing) between temperate and tropical breeds

At equilibrium two-thirds of the hybrid vigour of the F_1 is retained. In alternate generations progeny will be two-thirds tropical or two-thirds temperate in their genotype. This system is very flexible; if more than an average of 50 percent temperate blood is needed then there can be two generations of crossing with the temperate breed alternating with one cross with the tropical breed. In this case at equilibrium succeeding generations will have 3/7, 5/7 and 6/7 of the temperate blood (Figure 2). Or the tropical breed can be replaced by a halfbred - either a new intermediate breed or a half-bred sire.

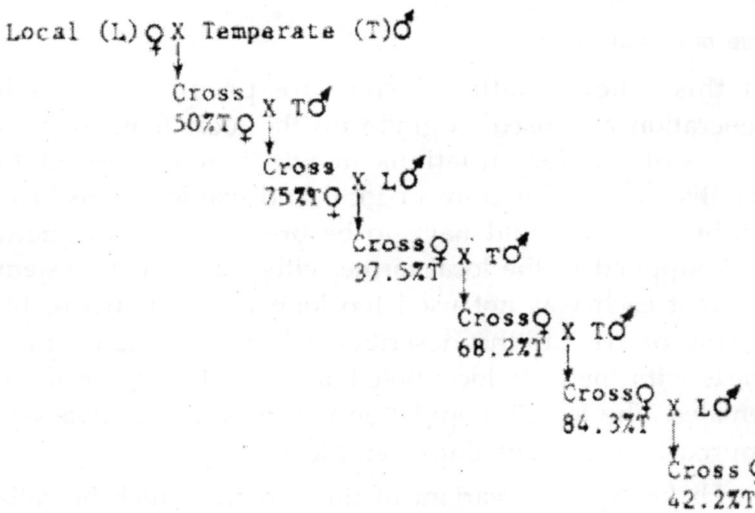

Figure 2. Rotational crossing between temperate and tropical breeds

A similar result can be achieved by using three different breeds in the crossing. It then becomes rotational crossing and 87 percent of the maximum heterosis is retained. This system is also flexible in the face of changing market demand - the type of crossing breed can be changed immediately according to the product needed. Systematic crossbreeding systems need a source of crossing sires. The tropical sires should come from an improved local breed under selection. The temperate sires should also come from nucleus herds selected in the local environment rather than from selection programmes overseas in view of the phenomenon of genotype x environment interaction.

Mixture of Terminal Crossing and Rotational Crossing

A meat sire can be used on the crossbred females which emerge in a crisscrossing or rotational crossing system. It is terminal because the offspring of this sire are not used for breeding but are all slaughtered for meat.

Use of crossbred sires

In this scheme halfbred sires are produced in each generation and used to grade up the local females. It is thus suitable for situations in which it is desired to stabilise a 50:50 mixture of local and exotic strains. The halfbred sires would have to be bred in a special herd and supplied to the local farmer either on a loan system (so that each was not used too long in one herd) or by means of AI. Lindhé describes a breeding plan which starts with the pure local breed and uses F_1 sires on these females. The female population will in a few generations approach 50 percent European blood.

Hickman has a variant of this scheme which he calls the "repeated hybrid male cross". The first cross on to the local females is by a pure European sire and the specially bred F_1 bulls are then used on these F_1 cows. The population is then 50 percent European from the first generation. The use of crossbred sires (or a new intermediate breed) is the only system possible in a herd too small to maintain two bulls for crisscrossing or with no access to AI. This scheme has the flexibility of the systematic crossing programmes and the simplicity (one type of sire only) of the gene-pool concept.

Indeed it can easily lead to the formation of a new breed if at any time it is decided to use sires bred within the population instead of using the specially bred F_1 bulls. It must be emphasised once more that all these schemes require a source of temperate sires which should preferably be selected in an environment similar to that in which they are to be used. They also require a selection programme for the indigenous breed which is used in the crossing programme.

Choosing Breeding System

The various options available in choosing a breeding system are shown diagrammatically in Figure 3. This includes the questions we have already discussed, i.e. whether to select in the local population or to cross with an exotic breed, which exotic to use, whether to grade up completely to the exotic or whether to maintain a crossbred population. However, one point is not made clear in this diagram. In conditions too adverse to maintain any proportion of European blood, the best breeding system may be to cross two tropically adapted breeds.

The gains from crossbreeding arise from:

i. complementarity, i.e. the combination of the adaptation of the tropical breed with the productivity of the improved temperate breed, and
ii. heterosis or hybrid vigour which, whatever its genetic explanation, demands a heterozygous genotype.

The first effect is additively genetic; the second is non-additive. From a genetic viewpoint the optimal method to maintain a crossbred population depends on the amount of heterosis exhibited in the cross. If heterosis is important then one of the systematic crossing systems should be used. The commercial crossing system exploits 100 percent of the heterosis, the 3-breed rotational cross 87 percent and the criss-cross 67 percent. A new breed, on the other hand, retains only half the heterosis of the first cross and this is the genetically preferable system if complementarity is important rather than heterosis.

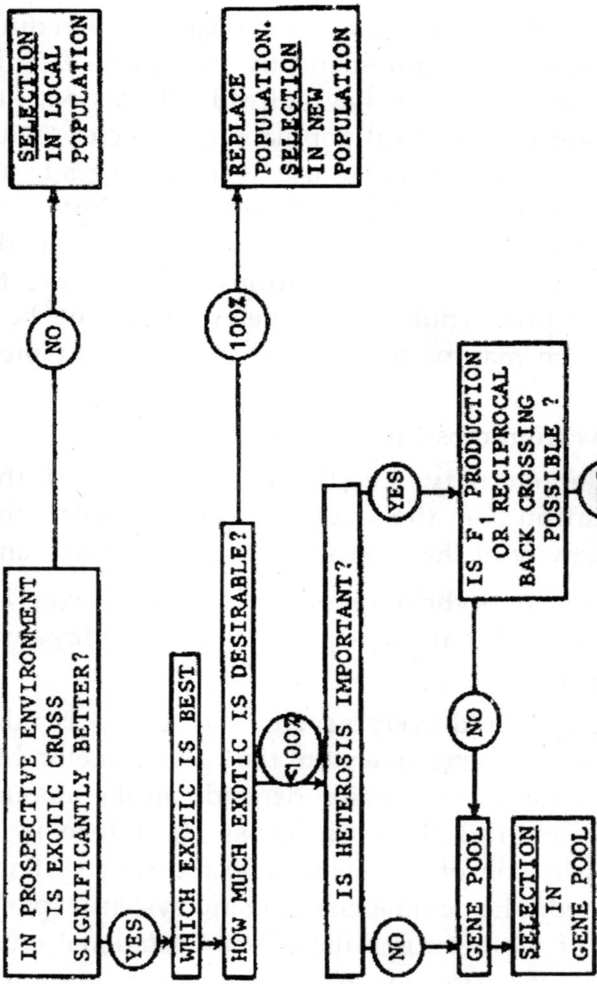

Figure 3 Strategic options involving crossbreeding and selection

In theory heterosis should be greater for those characters which are related to "fitness" and have therefore been under the influence of natural selection for a long time and now have a low heritability. It is also expected to be higher in a stressful environment. In practice it is necessary to test each cross in each situation. Cunningham has presented a model which would explain the common experience that heterosis is greater in a stressful than in a favourable environment. His model is shown in Figure 4.

HET = Heterosis

ADD = Additive effect

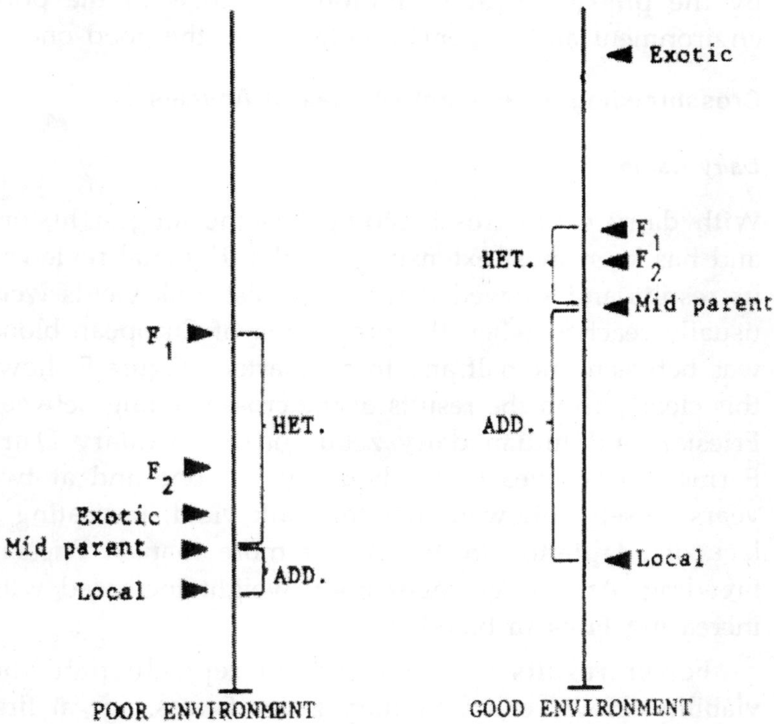

Figure 4. A model for differential heterosis and additive effects of two breeds in good and poor environments.

In a very unfavourable environment the local breed and the improved exotic breed may not differ greatly in production. The F_1 often shows a high performance, indicating heterosis. In a favourable environment, on the other hand, the improved breed will yield nearer to its genetic potential but the local breed does not significantly improve in performance. The F_1 is improved by an intermediate amount which may put it only slightly above the mid-parent indicating very modest heterosis. On this model the difference between the F_1 and the local strain is largely due to heterosis in the poor environment and to the additive genetic difference between the breeds in the favourable environment. This might be explained by the physiological limitations of stress in the poor environment and of genetic potential in the good one.

Crossbreeding in Different Classes of Animals

Dairy Cattle

With dairy cattle crossbreeding has the longest history and has been most extensive in India. Katpatal reviewed its results and showed that the highest milk yields were usually reached when the proportion of European blood was between one-half and three-quarters. Figure 5 shows this clearly from the results of the crossbreeding between Friesian and Indian dairy zebus on the Military Dairy Farms. The curves for body weight at one and at two years closely followed that for milk yield, indicating a lack of adaptation in animals of more than ¾ Friesian breeding. At younger ages body weight increased with increasing Friesian blood.

Fewer results are reported on reproduction and viability. However on Military Dairy Farms, age at first calving, calving interval and service period were generally lowest in the half breds and 5/8 breds. Striking results are quoted on mortality to first calving: it was 28 percent in purebred Sahiwals, and 47 percent in 31/32 bred Friesians; in half-breds and 5/8 breds on the other hand it was only 3-4 percent.

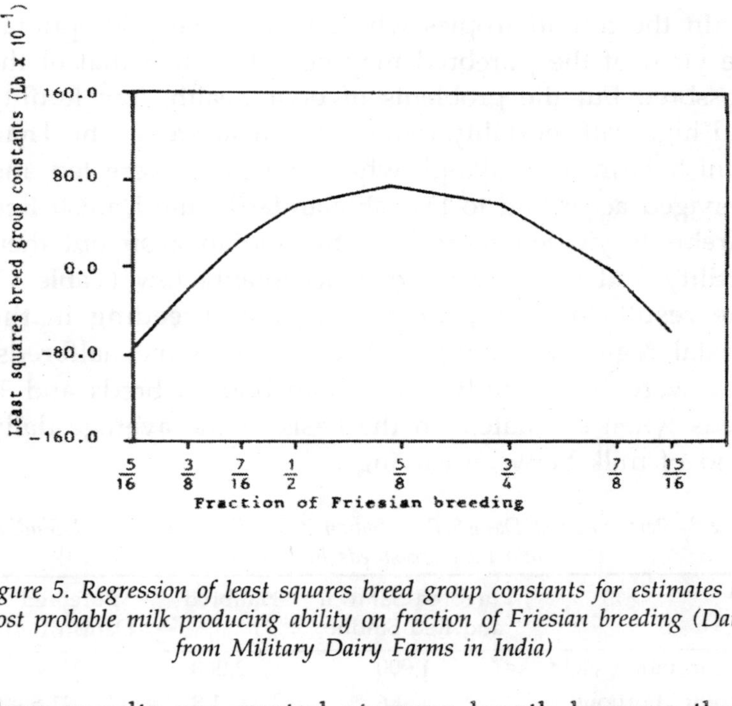

Figure 5. Regression of least squares breed group constants for estimates of most probable milk producing ability on fraction of Friesian breeding (Data from Military Dairy Farms in India)

These results are quoted at some length because they refer to several breeds, to large numbers of animals, to a variety of environments many of which are north of the tropic of Cancer, and to well managed government farms. They should therefore have some general application. Indeed in the semi-arid and coastal areas of Kenya and mainland Tanzania it is now clearly established that the Sahiwal x European dairy cross outyields both purebreds and has a better reproductive record and lower mortality. The optimum proportion of European blood varies according to management between onehalf and three-quarters. Even in the high potential areas of Kenya under conditions of poor feeding and management the crossbred often has an advantage in reproductive rate (including fertility, abortions and calf mortality) and may even outyield the purebred European.

In the humid tropics where management is optimal, the yield of the purebred may be better than that of the crossbred but the problems of poor health, low fertility and high calf mortality remain. For instance, at the Thai-Danish farm in Thailand, where the cattle were fed and managed according to Danish standards, the Danish Red purebreds yielded more than the Indian zebu but their fertility and viability were unacceptably low (Table 1). The results of European x zebu crossbreeding in the coastal region of Tanzania also favoured the halfbreds; they were significantly better than both ¼ breds and ¾ breds when evaluated on the basis of the average daily yield of milk between calvings.

Table 1. Performace of Danish Red, Indian Zebu (Sahiwal and Red Sindhi) and their crossbreds in Thailand

	Purebred Sahiwal and Red Sindhi	Halfbred	Purebred Danish Red
First lactation yield (kg)	1,000	2,000	2,300
Percent abortions	4.6	1.8	21.5
Percent stillbirths	6.6	4.4	5.1
Mortality to 6 months (%)	15.5	5.9	7.3
Mortality from 6 months to calving (%)	5.4	0	23.7
Calving interval (days)	467	443	525

There is now good evidence for hybrid vigour when zebu and temperate cattle are crossed in the humid tropics. It is most often revealed by the decline in yield and reproductive rate as the crossing proceeds from F_1 to F_2, from F_1 to backcross or from the ¾ bred to the 7/8 bred. If the improvement in yield were entirely due to additive effect of genes the yield should increase as the proportion of European blood increased and the F_2 should have the same yield as the F_1.

This decline from F_1 to F_2 is shown clearly in the figures that Katpatal quotes from India. It is particularly striking in Sri Lanka - Table 2.

Table 2. Hybrid vigour in dairy cattle crosses in Sri Lanka as shown by decline form F_1 to F_2

	Milk yield(kg)	Calving interval(days)
Friesian x Sinhala		
F_1	1,573	393
F_2	987	448
Jersey x Sinhala		
F_1	1,215	370
F_2	809	412
Jersey x Sindhi		
F_1	1,929	368
F_2	1,115	430

This same decline will occur as soon as the first crosses are interbred in the second stage of forming a new breed. It has not prevented work on the formation of new breeds, but it would suggest that in the humid tropics a systematic crossbreeding system would be genetically more efficient. A reformation of the F_1 every generation is not applicable to dairy cattle; nor is a two-stage crossing system. The generation interval is too long and it is the breeding animal which is important, not its progeny for slaughter.

In spite of these shortcomings Madalena reports that in the north of Minais Gerais, Brazil, halfbred zebu heifers are produced for sale to more specialised dairying regions. Because of the short productive life of the zebu cows and the high mortality of their calves it is calculated that only 30 percent of the zebu herd could be used to produce F_1 calves. It would probably be better to use these herds for crossing with European beef bulls.

On the other hand a crisscrossing or rotational crossing system would be ideal and easy to apply, especially in a zebu x European breed cross. The semen from the two breeds could be distributed by an AI system; the hump of the 2/3 zebu cows would indicate the need for European semen and the zebu semen would be used on the humpless cows. If AI is not available it would be best to separate the two types of cows into different herds and to run a zebu bull with one and a European bull with the other.

Apart from its genetic merits this breeding scheme has several advantages. The total resources of the farm can be used for keeping cows and breeding their replacements. Bulls are not reared; all male calves are slaughtered and bulls (or semen) are brought in as necessary. The continuing genetic improvement stems from these bulls - no recording is needed in the herd except to breed replacement cows from the highest yielders and best breeders. They must be chosen equally from the two types, even though the 2/3 exotic cows may give more milk. Otherwise the proportion of the two types in the herd will become unbalanced. It would be desirable to start the crisscrossing by backcrossing half the F_1 cows to the zebu bull and half to the exotic bull. This would enable the two types to be kept in equilibrium from the start.

There may initially be some resistance to such a scheme. After the boost obtained by one generation of crossing with the dairy breed, farmers may be reluctant to return to the zebu bull. However, in some places this crossing system has been gradually developed by the farmers' own experience - for instance in Mexico and in Brazil. Meyn and Wilkins recommend a crisscross between Sahiwal and European dairy breeds for dairy ranching in the semi-arid and coastal regions of Kenya.

If more European blood is needed then the zebu bull is replaced by a halfbred or by a bull from a new halfbred breed. Madalena suggests that in Brazil the new Pitangueiras breed could be tried experimentally in this role. The extra European blood could also be introduced by using two European breeds to one zebu in a rotational cross. These might well be the Friesian and the Jersey in order to exploit the features of both the two outstanding dairy breeds. In alternate generations the cows would then have the following composition:

1. 4/7 Friesian 2/7 Jersey 1/7 zebu x zebu bull
2. 2/7 Friesian 1/7 Jersey 4/7 zebu x Jersey bull
3. 1/7 Friesian 4/7 Jersey 2/7 zebu x Friesian bull

It will be seen that the bull is used of the breed with the lowest proportion of blood in the cows. Since cows of all three types will in fact be present in the herd at any one time some care will have to be taken that accurate breeding records are kept and the right breed of bull is used for each cow. This demands either AI or segregation of the cows into three separate herds. Each cow should be marked with its breed type.

In the most adverse conditions the local breed must be retained but it might be worth trying a cross with another tropical dairy breed. In Latin America the cross might be between a dairy zebu and a dairy Criollo, in West Africa between a dairy zebu and the N'Dama, in the eastern Mediterranean between a dairy zebu and the

Shami breed. In Kenya the initial improvement of the local zebu was with Sahiwal bulls.

All these crossing systems are fairly sophisticated and some are highly so. When the first cross has been successful and the question arises, what to do next, the simplest solution is to use crossbred bulls and it is also satisfactory genetically. Many development projects which started with purebred dairy bulls are now using halfbreds on the crossbred cows e.g. in India, Ethiopia.

Beef Cattle

For beef cattle it has been clearly demonstrated that there are many pastoral areas in Africa, Australia and tropical America, where the European crossbred performs better than the local adapted cattle or the purebred European beef breed. For instance, Sacker et al. showed that Red Poll x Boran crossbred calves were 26-30 percent heavier than the mean of their purebred parents at 3-12 months of age (Figure 6).

The most striking results have come from Queensland, Australia, where the zebu x European crossbred has excelled the purebred European breed in growth rate, fertility and viability. In these experiments the zebus were Africanders and Brahmans and the European breed was a Shorthorn x Hereford cross. Most of the experimental work compared the crossbreds only with the European parent. Smaller trials showed them to be superior also to the zebu parental breeds except in postweaning and adult mortality.

Since beef cattle are more exposed to the environment than are dairy cattle, crossbreeding for beef is advantageous even in climates where the best dairy breed would be a pure European. For instance in the southern United States cows which are crosses between the

American Brahman and European beef breeds exceed the average of purebred controls by 27-46 percent when productivity is measured as weight of weaned calf per cow mated.

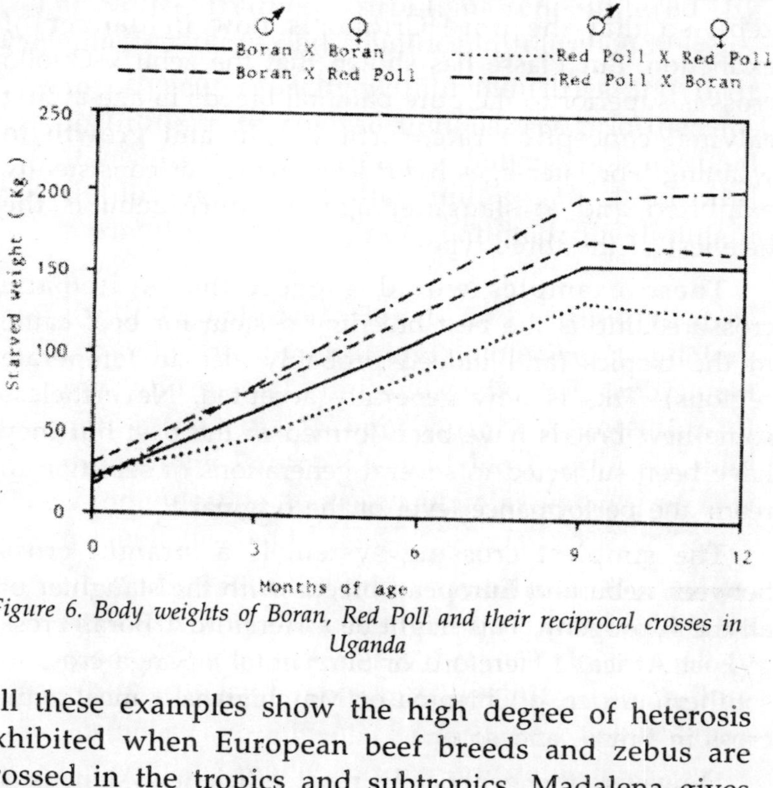

Figure 6. Body weights of Boran, Red Poll and their reciprocal crosses in Uganda

All these examples show the high degree of heterosis exhibited when European beef breeds and zebus are crossed in the tropics and subtropics. Madalena gives examples from Latin America of the improvement in growth rate and reproductive rate when the Criollo is crossed with the zebu or the zebu with a European beef breed. However, in few cases have both parental breeds and the cross (or better the two reciprocal crosses) all been included in the trial so that heterosis is not proved. This is an important point. If only the local breed and its cross with a temperate breed are available, the superiority of the cross should not be interpreted as a reason for continuing to grade to the imported breed. This point was overlooked in South America.

The initial improvement from crossing zebus on to the Criollo in the tropical areas has led to grading up to the zebu so that the pure Criollo is now in danger of extinction. But Plasse has shown that the zebu x Criollo cross is superior to the pure parental breeds in age at first calving, conception rate, birth weight and growth to weaning. For later ages hybrid vigour is not consistently exhibited and at slaughter age the pure zebu is the heaviest of the three types.

These examples would suggest that systematic crossbreeding is the best breeding system for beef cattle in the tropics (and indeed probably also in temperate regions). This is now generally accepted. Nevertheless some new breeds have been formed in the past but they have been subjected to several generations of selection to reach the performance level of the original F_1.

The simplest crossing system is a straight cross between zebu and European breeds with the slaughter of all the F_1 progeny. This might be a Hereford x Boran cross in East Africa, a Hereford or Simmental x Sanga cross in southern Africa, a Chianina or Marchigiana x meat zebu cross in Brazil, and so on.

However, the genetically most efficient system is a terminal cross; this exploits the hybrid vigour in the cows in such characters as fertility and mothering ability. It is a common system in Britain where the Galloway (a hill breed) is first crossed with a white Shorthorn to give the Blue Grey cow and the terminal cross is with a Hereford (or one of the large European breeds). The first-cross steers and all the terminal cross - steers and heifers - are reared for beef. The first cross should be with a small breed (for low maintenance costs) with high fertility. In

Latin America the first cross might be Criollo x zebu and the terminal cross with a large European beef breed.

In Africa the first cross on to the local zebu or Sanga breed might be an improved zebu (Boran or Africander or American Brahman) or an intermediate breed (Santa Gertrudis) or a small European breed (Jersey or Aberdeen-Angus) according to the environment. The terminal cross would be with a Hereford or one of the large European beef breeds (Charolais or Simmental) according to the feeding level which can be provided for the progeny.

In many situations crisscrossing or rotational crossing is to be preferred to terminal crossing. It has the great advantage that young females are not slaughtered for meat. It is thus appropriate in countries where there is a legal ban on the slaughter of young females, and also in circumstances where the aim is to increase herd numbers. A crisscrossing system has been in operation among beef cattle breeders in southern Africa for many years on an ad hoc basis. It usually involved crosses between South Devon or Hereford and Mashona in Zimbabwe and between Hereford and Africander in South Africa. Because it was based on the farmers' experience rather than on genetic theory they tend to describe it in a defensive, almost guilty, way which is entirely unnecessary since it is genetically sound. Based on breed availability one might suggest a Hereford x Boran crisscross in East Africa and a zebu x European beef cross in tropical America.

For harsher conditions a cross between two unrelated tropical breeds would probably give the best results, for instance, Brahman x Tswana in Botswana, zebu x Criollo in Latin America. If three tropical breeds are available a rotational cross can be made. In Costa Rica the advantage of crossing Santa Gertrudis, Brahman and Criollo in all possible ways has been demonstrated but this experimental result has not yet developed into a commercial breeding system.

The terminal crossing and crisscrossing systems can be combined by using a crisscross between two tropical breeds to produce the cow herd. They will combine tropical adaptation and hybrid vigour and so should be hardy, fertile, long lived and good mothers. Onto a proportion of these crossbred cows which are not needed for breeding replacements is put a bull of a European beef breed. The progeny should be good beef animals but they will need better conditions than their mothers and adequate feeding to develop their potential. Both males and females will be reared for beef.

All these crossbreeding systems need careful identification of cows so that the right type of bull can be used on each type of cow. Some are too sophisticated for general use. The important thing is to use a crossbred cow. The additional advantage from more refined systems are not often worth the extra trouble. Progeny from F_1 dams mated to a third breed were only slightly superior to backcross progeny, and progeny from an established crisscross or rotational crossing system were only slightly inferior. As expected on theoretical grounds inter se mating of F_1 bulls and cows leads to a considerable decrease in hybrid vigour. They advocate the crisscross breeding system-especially for zebu x European breeding systems in environments comparable to the Gulf Coast of America. It is simple in having only two breeding groups; all the females are crossbreds; there is scope for selection among the females; and heterosis is almost as high as in the system of breeding from F_1 females (terminal crossing).

Sheep Crossbreeding

Most of the sheep crossbreeding in the tropics has been in the form of gradingup or the formation of new breeds. There have been few successful systematic crossing programmes. This is not because heterosis has been shown to be absentin fact there have been few experiments designed to detect it and little data on which it could be studied. Sheep in the dry tropics are kept extensively on the range and under these conditions it is not possible to cross with an unadapted breed however productive it may be in its home environment. In the wet tropics sheep are kept on a subsistence basis by smallholders and in small numbers. They could not easily support the sophistication of even the simplest crossing scheme.

Wool

Merinos have been successfully introduced into many subtropical areas (Australia, South Africa, Uruguay, Mexico) but the only areas within the tropics where they flourish are Queensland and the north of Western Australia on the one hand, and the highlands of Kenya on the other. Little crossbreeding was involved in these introductions. In Egypt (which is not strictly tropical) there have been two attempts with Merinos. The first was in the middle of the 19th century but failed owing to the poor care the sheep received. The second was in the 1950s and 1960s. Various breeds of European Merino were imported but they suffered very much from pneumonia due to dust. Crosses onto the local coarse-wooled breeds were intermediate in fleece weight and wool quality but had a highly variable fibre diameter and coloured patches on the body. However, they showed hybrid vigour in lamb viability and growth rate.

In West Africa Merino crossing has not been a success. Grading of local hair breeds in Chad during the 1930s was discontinued owing to a lack of market for wool. Undertaking a similar cross in northern Nigeria, Burns found that the F fleece was excessively kempy. She argued that another breed with "central checking" might be more successful than the Merino in reducing kemp and introduced a Wensleydale cross for this purpose. Incidentally little trace now remains of the crossbred animals produced in Katsina.

All the above trials have been rather ad hoc and experimental. None led to any permanent results. It is only in India that there has been a large-scale development programme for sheep which has included crossbreeding to improve wool production. In Kashmir and other Himalayan regions the local breeds have been successfully graded to American Rambouillet and various Russian finewool breeds to produce new breeds. Elsewhere the plan was to keep Corriedales in state farms and use them to cross on to farm flocks.

Tremendous difficulties were encountered in the early years in keeping the Corriedales alive and reproducing. There has thus been a delay in starting the comparison of crossbreds with indigenous sheep and there has not been any evaluation of the crossbred wool and its uses. These steps are an essential preliminary to any large-scale crossbreeding project.

For the crossbreeding an initial 5 years is allowed for the testing of the various possible crossbreds. The rate of spread of exotic genes is then calculated based on the distribution of crossbred rams which at first are 50 percent rising to 75 percent exotic blood. In 10 years, even starting with purebred rams, the level of exotic genes in the general population will be only 36 percent. Under these conditions it is calculated that the exotic breed must be 60 percent superior to the indigenous for crossing to produce the same genetic improvement in the population as selection in a period of 10 years.

Meat

One would expect that meat production of tropical sheep could be improved by some sort of systematic crossing system, as with beef cattle, based on the local breed and on an improved mutton breed. No such programmes have developed. There have been sporadic crossbreeding trials but none has left permanent results. Even in subtropical countries such as Iran and Afghanistan, Yalçin concluded that the results of experimental crossbreeding had not been sufficiently encouraging for it to be extended to the much poorer environmental conditions in the field. In addition to the prevailing unfavourable nutritional and managemental conditions there is the difficulty of natural mating between thin-tailed foreign rams and fattailed native ewes.

In the tropics results have been equally discouraging. Wool and meat-wool breeds have been imported into Java off and on since 1864 but no trace of these early imports remains. More recent trials with Dormers and Suffmers from Australia have not proceeded far but they are confined to well managed experimental herds in the highlands. European sheep were also imported at various times into the Caribbean islands but were not used systematically. They account for the partially woolled animals which appear among the generally woolless population.

European breeds and Indian woolled breeds did not thrive in Sri Lanka. The best results were obtained with hair breeds from south India (especially the Bannur) under extensive conditions and with Dorset x local hair sheep for intensive husbandry. In other places haphazard crossing has led to the formation of new intermediate populations which are now recognized as breeds. For instance the sheep of the Nilgiri hills in southern India, which have a short fine fleece, are derived from a cross of the local coarse-wooled or hair sheep with Merino and Southdown in the early 19th century.

A more recent new breed is the Santa Inês of north-east Brazil. This developed since the 1940s by crossing the Bergamasca breed of Italy with the local hair sheep. It gave them greater size without significantly reducing their adaptation or clothing them with wool. It was formed by chance but is now accepted and is being selected as a new breed. In South Africa many crosses were made between various European wool breeds and local hair sheep in order to combine their mutual advantages. The one which was most successful was the Dorset Horn x Blackhead Persian.

The crossing programme was begun in the 1930s and a breed society for Dorper sheep was formed in 1950. Its breeding is 50:50 but there have been many generations of selection to fix its characteristics. The Dorper inherits the black head, hornlessness, and adaptation of the Blackhead Persian but it lacks its fat deposits on rump, tail and brisket. It has the mutton conformation and growth rate of the Dorset. The coat is a mixture of wool and hair but selection is against excess of wool. This breed has been used as an improver in other parts of southern and eastern Africa, especially in Kenya. Similar efforts are being made elsewhere.

In Cuba, the government policy is to form a new breed by crossing the local hair sheep with the Suffolk and trials are in progress to determine the optimum proportion of Suffolk blood. This policy may be satisfactory for State farms but probably under ordinary levels of feeding and management it is the pure hair breed which will perform best. Too little work has been done on the crossing of two adapted breeds to exploit either hybrid vigour or complentarity.

There are some indications that in the dry subtropics the local coarse wool breed can be improved by crossing with a similar breed from elsewhere. Thus in Libya the Barbary has been improved in size, weight and fleece weight, by crossing with the White Karaman from Turkey. In Egypt hybrid vigour is exhibited when the Ausimi and Rahmani are crossed. In Iran the Shal breed has been improved by crossing with the Israeli Awassi: heterosis was exhibited in lambing rate, litter size, milk yield and growth rate.

There has also been some crossbreeding among hair sheep breeds in the tropics. The chief improver breed used has been the Blackhead Persian which has left extensive traces in southern and eastern Africa and in the Caribbean. In Brazil a new breed has been developed - the Brazilian Somali - which retains the black head but has almost lost the fat rump. Crossing with the woolled Criollo has introduced a certain amount of wool.

That these crossbreds have persisted in the absence of any special breeding programmes indicates their thriftiness and suggests that more systematic effort should be put into mutual improvement of tropical breeds - and especially hair breeds - with the eventual aim of forming new breeds based on crossbred foundations. In this connection the prolific tropical breeds have a special role to play since, to improve meat production, increase in number of lambs may be more important than increase in growth rate.

The two most useful improver breeds would be the Barbados Blackbelly and the White Virgin Island sheep.

They may also be subject to improvement themselves by crossing with European mutton breeds. Indeed the high litter size of the Barbados Blackbelly may itself be the result of early crossing between the hair sheep introduced from Africa and a prolific breed from the Netherlands.

The White Virgin Island sheep has been used (in Maine, USA) to form an improved prolific hair breed by crossing with the Suffolk and Wiltshire Horn - namely the Katahdin. They are larger than the Virgin Island sheep but just as prolific (lambing rate up to 200 percent). Similar initiatives should be undertaken in the tropics using local sheep.

Goats

In the tropics the goat is the producer par excellence of milk and meat for home consumption. The small size of the flocks of subsistence farmers precludes a sophisticated crossing programme and the poor feeding and management are not suitable for the introduction of European blood. The most successful results have been obtained by the infusion of blood of improved tropical breeds. Above all the Jamnapari from India has been used to improve milk yield and size (and hence meat production) in several countries of southeast Asia.

In Malaysia and Indonesia the majority of goats now show evidence of Jamnapari blood in their size and lop ears as well, of course, as their milk production. The Jamnapari has also been used in the West Indies. The Anglo-Nubian has been used for upgrading indigenous stock in the West Indies, Mauritius, Malaysia and the Philippines.

Although these two breeds have also been used in Brazil the breed which has had the most influence in the northeast of that country is the Bhuj (from Gujarat, India).

It has introduced lop ears and larger size into the original Crioulo goats. However it has not been proved that its economic return is better considering the reduced hardiness which accompanies the crossing.

Crossing with Swiss breeds to improve milk production has been attempted in the Caribbean, East Africa, Malaysia, Fiji, Mauritius and Queensland. In West Malaysia the Toggenburg was the least satisfactory. Saanen and British Alpine crosses on to the local goats were superior to the local goats in milk production but for all round performance the Anglo-Nubian was the most successful improver. Among meat breeds only the Boer (of South Africa) has a reputation but it has in fact not been very successful in East Africa.

It can be concluded that with goats even more than with sheep crossing with European breeds must be undertaken only with the greatest caution. For both milk and meat production the need is for selection programmes to produce improved tropical breeds which can be used for grading up populations in which such a selection programme is not possible.

13

Breeding for Sustainable Agriculture

Plant breeders have greatly contributed to the development of high-yielding crop varieties to meet the food needs of a growing population, during the past three decades. Breeders have changed the morphology and physiology of crop plants, and incorporated desirable traits and resistant gene(s) into traditional varieties while stabilizing or increasing crop production. In most rice-growing countries, the primary breeding objective has been a high yield potential. The incorporation of the semidwarf gene into traditional tall, leafy rice varieties has made possible dramatic yield increases in the major rice growing countries (Table 1).

The semidwarf rice variety IR 36 and other IR lines have been intensively used in national breeding programs for the development of high-yielding varieties. These are now planted in 60% of the world's rice land. In Korea, Tongil-type varieties derived from crosses between IR lines and Japonica varieties give yields 30% higher than the leading traditional varieties.

High-yielding varieties have made a great contribution to the world's food supply, but they also have several major problems. The high yields of these varieties can only be attained with a high level of inputs, in particular heavy applications of fertilizer (Fig. 1). This has led to problems associated with pest outbreaks in certain areas, while increased rice production has resulted in lower rice prices.

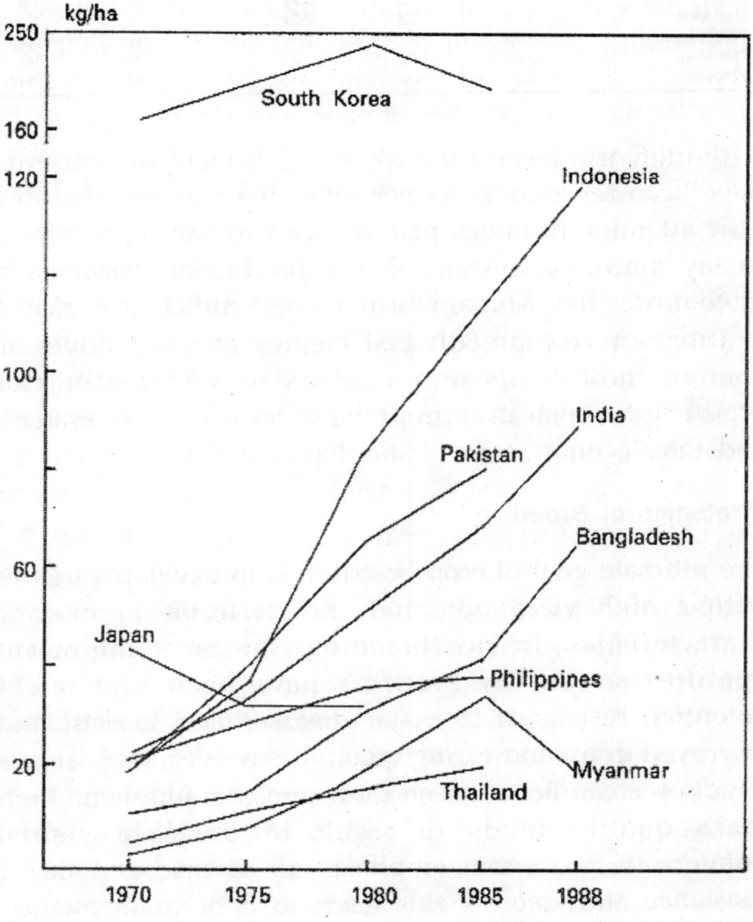

Figure 1. Consumption of fertilizer (N, P_2O_5, K_2O) applied to rice

Table 1 Rough rice yield in selected Asian countries, World Rice Statistics,

Country	Yield ton/ha			% increase over 1970	
	1970	1977	1987	in 1977	in 1987
Bangladesh	1.69	1.94	2.24	115	133
China	3.42	3.64	5.41	106	158
India	1.68	1.96	2.29	117	136
Indonesia	2.38	2.79	4.04	117	170
Japan	5.63	6.17	6.19	110	110
South Korea	4.55	6.79	6.02	149	132
Myanmar	1.70	1.95	3.04	115	179
Philippines	1.75	2.00	2.62	114	150
Thailand	2.02	1.59	2.02	79	100
Vietnam	2.15	1.94	2.70	90	126

Although numerous improved high-yielding varieties have been released to farmers over the past few decades, little attention has been paid in rice varietal improvement to any concept of sustainable rice production. Particularly in countries like Korea, which are self-sufficient in rice, it is time for rice breeding strategies and objectives to change their focus and emphasize sustainable rice production which is appropriate to local farming systems and their economic and social background.

Strategies of Breeding

The ultimate goal of crop breeding is to develop varieties with a high yield potential and desirable agronomic characteristics. In rice breeding, the most important qualities sought by breeders have been high yield potential; resistance to major diseases and insects; and improved grain and eating quality. However, there seems to be some conflict between these aims. Emphasis on high grain quality tends to result in unstable yields. Conversely, too much emphasis on disease and insect resistance and stable yields leads to poor grain quality. Hence, breeding efforts should concentrate on varieties with the potential to minimize yield losses under unfavourable conditions, and to maximize yields when conditions are favourable. The following breeding approaches should be emphasized in producing varieties for sustainable rice production.

— High-yield potential under low inputs.

— _ Heterotic F_1 hybrid

— _ New plant type

— Premium grain and eating quality to meet consumer demand, and to provide grain suitable for processing.

— More genetic diversity.

— Durable host resistance to major diseases and insects.

— Wider range of growth duration for various purposes.

— Proper levels of tolerance to environmental and climatic stresses in specific areas.

High Yielding Crop Varieties

For sustainable agricultural production, attention must focus on the development of high-yielding crop varieties which have premium quality under low inputs. At present, breeders are examining a wide range of genetic resources towards this end. Experiments have indicated that average yields of current rice varieties given no nitrogen application are only 70% of those given an optimum N application of 150 kg N/ha (Fig. 2). However, F_1 hybrids are one possible way in which we may achieve yield increases with less applied nitrogen in the near future.

A study of the growth of rice hybrids given five levels of nitrogen fertilizer showed that positive heterobetiosis in plant weight increased linearly from no nitrogen to 150 kg of nitrogen per hectare (Fig. 3). The highest heterobeltiosis in plant weight was found when no nitrogen was applied, followed by 75 kg/ha and 150 kg/ha of nitrogen. In a comparative study of nitrogen response in F_1 hybrids and conventional semidwarf varieties, F1 hybrids were observed to have higher yields when no nitrogen was applied (Fig. 4). This suggests that hybrids attain a higher yield at lower nitrogen applications than do conventional semidwarf varieties.

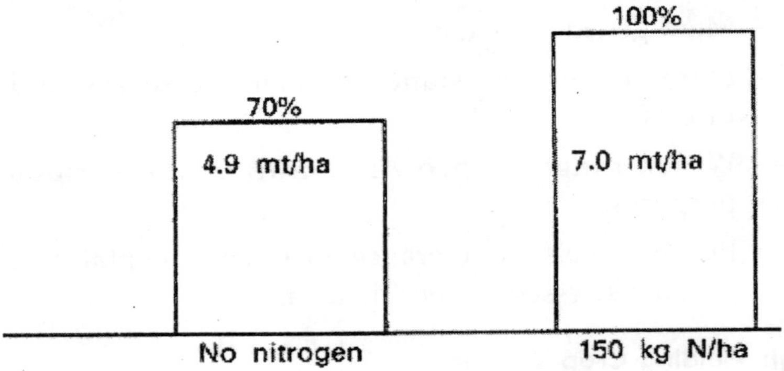

Figure 2. Average grain yield production under no nitrogen and 150 kg/ha

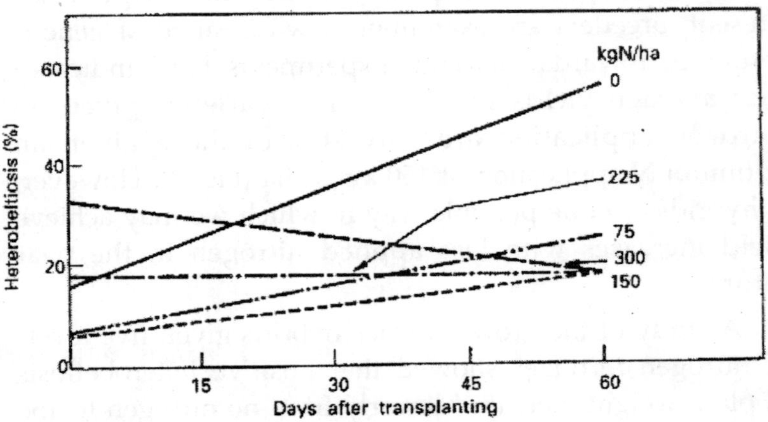

Figure 3. Change in heterobeltiosis on dry weight 60 days after

In Mainland China, about 14 million ha are currently planted in hybrid rices, but the use of hybrids in other areas would depend on whether seed production was economical, and on whether grain quality was found acceptable by consumers. Recently, photoperiod-sensitive (PGMS) and thermosensitive male sterility (TGMS) systems are giving promising results in seed production. It has been suggested that one approach to achieving high yields with low inputs may be to emphasize varieties with the characteristics of staggered heading during the flowering stage.

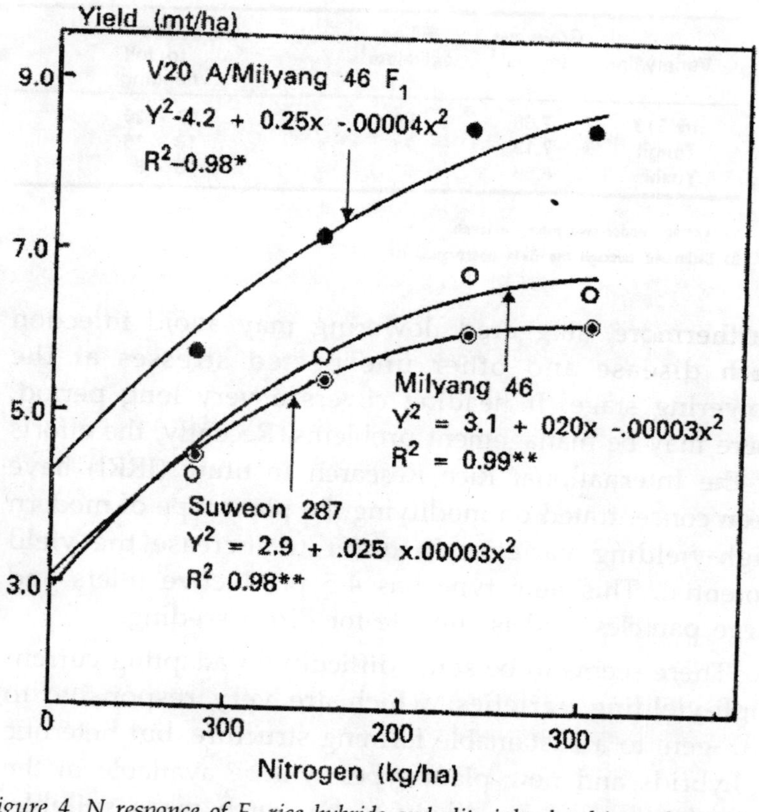

Figure 4. N responce of F_1 rice hybrids and elite inbred cultivars for yield

A major barrier at present to even higher yields is that varieties with a large number of spikelets per panicle tend to have a lower percentage of filled spikelets, caused by nutritional competition between grains. Staggered heading of tillers over a prolonged period, rather than simultaneous heading within a short period, may overcome nutritional competition between grains caused by an insufficient supply of carbohydrates as they simultaneously develop (Table 2).

Table 2 Comparison of grain yield and filled spikelet percentage, with staggered heading of three Tongil-type varieties/lines

Variety/line	Grain yield (mt/ha)[1]	Filled spikelets[1] (%)	Days from initial to full heading[2]
Iri 313	7.68	78	17 - 20
Tongil	7.13	71	13 - 16
Yushin	6.58	51	6 - 8

1) Average under two nitrogen levels

2) Estimated through the field observation

Furthermore, staggered flowering may avoid infection with disease and other unexpected stresses at the flowering stage. If heading covers a very long period, there may be management problems. Recently, the efforts of the International Rice Research Institute (IRRI) have been concentrated on modifying the plant type of modern high-yielding varieties in order to increase the yield potential. This new type has 4-5 productive tillers and large panicles, and is suitable for direct seeding.

There seems to be some difficulty in adapting current high-yielding varieties, which are very responsive to nitrogen, to a sustainable farming structure, but heterotic F_1 hybrids and new plant types will be available in the near future and are likely to be much more suitable.

Furthermore, high-yielding varieties can be expected to increase farmers' incomes, since they will permit more diversified cropping systems as less land is required for rice.

Insects and Disease Resistance

Needless to say, host resistance to various biotic stresses is a very important aspect of high yields, and can be expected to play a significant role in sustainable rice production. Intensive efforts have been made over the years to incorporate resistant gene(s) into improved varieties, and there are now numerous varieties resistant to rice blast, bacterial blight, various virus diseases, and planthoppers. Some varieties possess multiple resistance to diseases and insects, but resistance in most varieties is controlled by a single gene and lasts only a few years, after which they become susceptible to serious disease or pest outbreaks.

For example, most Tongil-type varieties grown widely in Korea during the 1970s had a high level of resistance to rice blast disease, but in 1977 large areas planted in these varieties were severely damaged by blast. In contrast, varieties with the Xa 4 gene resistant to bacterial wilt have been grown in the Philippines for the last 15 years, and continue to be resistant. If a resistant variety becomes susceptible, a new variety with different gene(s) for resistance should be waiting ready for distribution to farmers. However, it is difficult to predict in advance which variety will suddenly need to be replaced, and it is difficult to release a new variety within a short period.

Furthermore, there may not be a wide enough range of genetic resources to provide a new variety with different genes. It is extremely difficult to identify polygenic resistance and incorporate it into improved germplasm. Current studies on host resistance to crops emphasize the durability of resistance. Polygenic traits rather than absolute resistance would be preferable in sustainable agricultural production. Some breeders believe that resistance to pests or other stresses is linked with low yield and poor grain quality.

Grain Quality

Improvements in rice quality are very important in meeting the demands of consumers for healthy, high-quality food. Most rice breeders feel that improved high-yielding varieties in recent years have premium quality in terms of grain size, shape, appearance and palatability. However, consumers are still not satisfied with the grain quality of these varieties. There are many difficulties in evaluating consumer preferences, which vary widely from country to country and from person to person. However, consumers generally feel that even the best quality grain of improved high-yielding varieties is not as good as the best quality traditional varieties.

Many traditional varieties in both the tropics and the temperate zone have excellent cooking and eating quality, but a low grain yield. For many years, breeders have focused their attention on quality improvement, but without success. There seems to be some unknown genetic barrier to incorporating this trait into high-yielding varieties. At present, it is not possible to be very optimistic that we shall be able to develop high-yielding varieties with quality as good as that of traditional varieties, but our efforts continue, as does our work to develop high-yielding varieties with various grain characteristics suitable for processing.

High Quality Rice Breeding with Aromatic Rice Cultivar

The advent of various biotechnological techniques in

recent years has offered new tools for developing high yielding rice varieties possessing good grain characteristics, such as aroma, to boost rice export and meet domestic demand. Among these tools, anther culture is emerging as an effective technique in rice breeding, particularly when applied to potential hybrid.

Used as Parents

Four aromatic varieties were used as parents: DS 20, IR 68144, Kloong Luang 1, and Suphan Buri. Eight F_1 hybrids were selected for anther culture. N6 or MS were used as basic medium for anther culture. Panicles were collected prior to emergence from flag leaf. Cold treatment of panicles was carried out at 10°C for 7 days in the incubator. After cold treatment, the spikelets were surface sterilized with alcohol and $HgCl_2$. Anthers were taken out aseptically from the spikelets and placed on the callus induction medium. The cultures were incubated in the dark at 20°C. When the calli emerging from the anther lobe grew, they were transferred to the regeneration medium and placed under continuous fluorescent light for plant regeneration. A0 plants were observed and collected to advance to A1 generation and selected for A2 generation.

Promising Lines

Several promising lines from the hybrids Suphan Buri/ OM 1490 and Kloong Luang 1/CM 16-27 were evaluated for yield potential and quality traits. Some lines showing good performance have been tested in large scale in the Mekong Delta. In addition, more than 300 additional doubled haploid lines from other hybrids are under evaluation

14

Breeding Policy

Tropical animals increase in size and performance when well fed and protected from disease but their output does not reach that of the temperate genotypes. It is clear that a more elaborate approach is needed—a change in both genotype and environment simultaneously or a selection within the local breed under the local environmental conditions.

The selection of breeds for production must have begun very soon after the first domestication of their wild ancestors in the Middle East some 8,000–10,000 years ago. The high yield of the dairy cow, the fleece of the wooled sheep and the Angora goat, the high growth rate of the beef cattle breeds are all the result of selection. Cattle were used for draught long before they were milked. This means that milk was originally only a by-product (the major dairy animals were sheep and goats) and any selection for milk yield could only be subsidiary to their major role as working animals. If Europe had been compared with the tropics in the Middle Ages the differences in the function and productivity of their livestock would not have been nearly so striking as today.

Indeed it is only during the last 200 years starting in western Europe that specialised and high producing breeds of dairy and beef cattle have been formed. Before that time the typical cattle were triple-purpose—draught, milk and meat—and each locality had its own variety. Likewise sheep had to produce wool and meat and often milk as well. Each region and even each mountain valley had its own local breed.

Performance Recording

Performance recording is a necessary pre-requisite to effective decision making on breeding policy. But the choice of recording systems and type of records will differ with the other uses to which they are put, e.g. the monitoring of populations or herd level management. Nonetheless, they should all take into account the aggregate effects of the various traits of productive significance and should provide a characterisation of the environment under which the production is recorded.

National Level Decisions Recording

In many countries, a fair amount of information from institutional and commercial herds is already available which should be analysed before introducing new schemes for recording. Recording for the purpose of national decision making should take into consideration the ecological zones in which livestock are kept, the farming system, whether nomadic, settled, smallholder or intensive, the role of crops in the farming system, the breed or breeds used and the problems of disease, adaptation and survival in each breed/production system combination.

The initial step usually involves surveillance studies designed to obtain baseline data on productivity parameters and on the constraints to increased productivity. In their simplest form, they consist of interviews of chosen farmers conducted by selected personnel to obtain approximate parameters of birth, growth, production, mortality, etc. A representative sample of farms and animals in each combination of ecological zone, production system and breed should be chosen and, if any development project is being implemented, the response of traditional farmers to the developmental process should also be recorded.

The next step involves the regular monitoring of the performance of particular populations over a given period of time. The herds chosen should be representative of the different production systems and the monitoring should relate to herd structure, fertility, mortality and disposal, health and productivity. The parameters should, where possible, be related to feed inputs. An important aspect of the monitoring process should be the study of livestock development programmes. Such programmes are essentially large-scale experiments which cannot be reproduced within more conventional research programmes and can therefore provide invaluable information on the response to environments. Where such development programmes do not operate, it may be necessary to introduce improvements in some herds as part of the monitoring programme in order to provide this information.

Herd Level Recording

Performance recording of individual animals is usually associated with selection for genetic improvement. However, such recording may be even more important for management decisions. In the initial stages of introduction of a recording scheme, emphasis should be on immediate benefits that could be derived from management changes, e.g. more surviving progeny or an increase in saleable produce.

Cooperation between the farmer and the field recorder is a basic requirement for the successful operation of any recording scheme. This cooperation can only come through an understanding of the use of records. The recording scheme should therefore be capable of performing an educational role, not merely a service one.

Tropics Recording

The transfer of recording systems used in temperate countries to the tropics has in most cases proved impracticable. No standard system of recording can be devised that is suitable for all farming conditions. The level of recording that is possible depends on the kind of management adopted and the stage of development of the livestock industry. With large flocks reared extensively under nomadic conditions, the identification of animals with good fertility and longevity may be all that is possible. Recording is thus an evolutionary process and advances can be made only when there is progress in levels of husbandry.

At every stage, it is important that no more than the essential records are maintained. This is necessary not only because of inadequacies in the educational level of livestock farmers in the tropics but also due to the lack of qualified recorders to operate more sophisticated systems. A further reason is the inverse relationship between the rate of genetic progress and the number of traits for which selection is practised. Optimum deployment of recording systems therefore requires that it be limited to traits which can be measured easily and with a high degree of accuracy and which have an economic value.

In most tropical countries, large herds are kept in governmental and research institutes to serve as sources of superior male breeding stock. The recording done in these herds may involve extensive written records. In privately owned herds, the type of information that can be extracted depends on the farming system. For example, although it is futile to attempt recording milk yield or growth rate in nomadic herds, they can nevertheless provide valuable information on longevity and reproductive performance under conditions of nutritional stress and disease challenge.

Simple procedures (e.g. ear notching) can be adopted to place animals in different performance categories. This would aid in the selection of progeny that can then be subjected to further tests in testing stations or institutional herds. The chief advantage of nomadic herds is their large size which could allow a high selection intensity. In contrast, sedentary herds, despite their accessibility to recorders and the facility they provide for recording more traits, have the disadvantages associated with small size.

Any performance recording system in the tropics must satisfy a number of criteria if it is to be effective.

i) The system must be simple, i.e. involve little paper work for the farmer. It should not require major alterations in his farming routine.

ii) The traits should be economically important and the resulting products should be either marketable or consumed by the producer. Their number should be kept as low as possible. Information extracted should also aid in management.

iii) The system should be efficient in terms of time and cost.

iv) The records should make it possible to identify the best and poorest animals at the farm level and genetic differences between populations at the national level.

Dairy Cattle Recording

Practices of Milk Recording

The usual practice in milk recording is for milk and butterfat to be estimated from a limited number of measurements made at intervals throughout the lactation. The longer the interval between these measurements, the greater the error of the estimates is likely to be. When the interval is shortened, the expense of recording is increased, so that in practice a compromise is made between the needs of accuracy and economy.

Generally, milk recording in the developed countries has involved the weighing and sampling of the milk produced by individual cows on farms that are visited by official recorders at intervals of 3-4 weeks throughout the year. Morning and evening milk yields are recorded and composite samples of the two are taken for fat (and protein) determination. With the increasing cost of recording, there is a trend now towards the farmer doing his own recording. Monthly records may not be sufficiently elastic to meet the requirements of herd owners whose primary interest is in using the records for feeding and management. They are therefore encouraged to make entries of production at more frequent intervals—weekly or even daily.

Milk Recording in Tropics

Institutional herds in the tropics can adopt the same procedures of milk recording used in developed countries. Similar practices can also be used in large private herds. However, a large proportion of dairy cattle in the tropics are owned by small farmers and it is here that the problem of milk recording lies. The major constraints are the small average herd size, poor communication and lack of incentives for recording. Other possible reasons are inadequate appreciation of the value of progeny tested sires, low educational level of farmers and lack of qualified recorders.

Where AI services are relatively undeveloped, the value of a recording service may be questioned. However, even in such situations, recording can assist in the culling of unproductive females in addition to aiding in herd management. When introducing a milk recording scheme, certain incentives may have to be provided to encourage farmers to cooperate, e.g. free veterinary care, advice on management and feeding, low cost AI. It may even be necessary to pay a fee to farmers who cooperate in recording daughters of bulls used in progeny testing.

Official Recording

The main costs of milk recording are the recorder's salary and expenses. These tend to be high in the tropics because of the small size of herds, their scattered distribution and the need to include a large number of herds and cows to facilitate progeny testing. Different methods can be adopted to overcome the financial stress of recording. There could be a reduction in the number of visits paid to each farmer, so that the recorder may cover more herds and cows. Recorders may also be employed on a part-time basis and could undertake other work as well. Thus, AI personnel can be trained in recording work and may serve as part-time recorders. This has particular application in the tropics.

In countries with organised AI services and where AI personnel make daily rounds, it may be possible to arrange their routine so that they call on one or more farmers daily for milk recording. The technician need not do the milk recording himself but could provide the farmer with a recording device and on the following day collect it from him together with the milk records. It would of course be necessary for the technician to be present when the recording is done for the first time to demonstrate the procedure.

In situations such as the above, where it is not possible to have the services of official recorders except perhaps in a supervisory capacity, an owner sampling system can be introduced. Here, the farmer does the routine weighing himself, but periodic visits are made by the official recorder. The measuring equipment used by the farmer could be very simple. It may be plastic buckets or bottles pre-marked into ½ kg sections by the use of exact measuring equipment.

Milk Composition

In many tropical countries where milk is consumed primarily for its protein and not for fat, testing for butterfat content may not be necessary. However, bulk milk samples may be tested for fat content to assess the overall herd situation in instances where the price of milk is determined by its fat content. If it is necessary to know the fat composition of milk from individual cows, monthly testing of samples taken during the course of the lactation is adequate. Protein content is usually determined from fewer samples—four samples taken during a lactation being adequate. Semi-automatic machines for quick determination of protein content are now available which make it possible to analyse a large number of samples at a central laboratory. A preservative such as potassium dichromate should be added to the samples before despatch to the laboratory.

Samples for milk composition testing should preferably be taken by official recorders to ensure proper sampling procedures. Usually a composite sample proportional to the yield at the two milkings is taken for testing purposes. Testing for milk composition requires more organisation and a dependable communication system. Be that as it may, the importance of milk yield as a selection criterion far outweighs that of milk composition. Hence, non-availability of facilities to test for milk composition should not stand in the way of implementing a milk recording programme.

Monthly recording

i. *Bimonthly recording*: In this system, the recorder visits the farm every alternate month and records the yield obtained during a 24-hour period. The percentage of large errors of individual cow records in bimonthly recording is higher than with monthly testing but its level of accuracy is adequate for progeny testing purposes.

ii. *A.M.-P.M. sampling*: In this method, the morning milk is weighed one month, and the evening milk the next month. It is more accurate than bimonthly testing but less accurate than monthly testing. Its chief advantage is that it is cheaper than monthly testing but maintains the same frequency of visits to the farmers.

iii. *Sampling at particular stages of lactation*: The total yield estimated from monthly records is moderately correlated with the yield obtained on a single day during the third to the sixth month of lactation. This has led to the suggestion that a reasonably accurate progeny test of bull could be obtained from such records if large progeny groups are available. For example, a progeny test based on a single day's record of 100-150 daughters would be as accurate as a progeny test based on about 35 daughters with complete records. However, this system would be difficult to implement in herds calving year round since the cows will be in different stages of lactation when the recorder visits the farm. But if the farmers provide the data or if AI personnel are involved in the recording, it should prove workable.

The methods used in estimating total lactation yields depend on the recording system. The centering method of calculating yields is preferable to other methods in monthly recording. In this procedure, the test day yield is multiplied by the number of days in the period of which it is the centre, and these products are summed to give an estimated lactation yield. At the beginning and end of lactation, the test day will not necessarily be the centre day of the test period; the first test period runs from four days after calving to the beginning of the following period, and the last period runs from the end of the preceding period to the end of the lactation.

In general, fortnightly recordings are required for individual cow selection. Monthly recording gives reliable information for sire proving, but bimonthly or A.M.-P.M. methods may be used if a slight loss of accuracy is acceptable. If used as an aid in management, particularly in deciding on feeding levels, recording weekly or at more frequent intervals may be required.

Many zebu cows do not let their milk down unless the calf is allowed to suck. This problem is partially overcome by allowing the calf to suck for a very short period to induce milk let-down followed by hand milking. This process of stimulation of the dam by the calf and milking may be repeated if necessary until all the milk has been drawn. In using this method, the calf should be separated from the dam for the period intervening between P.M. and A.M. milking; otherwise the first record taken in the morning will be erroneous.

In most zebu breeds, milk yields and lactation length are strongly correlated. Lactation length can thus be used as an indirect selection trait in situations where yield is not measured, e.g. in nomadic herds. Age at first calving and calving intervals can be readily measured and are important indicators of adaptability to stress conditions.

Records on the Farm and Production

In most farm situations in the tropics, the individual cow record is the only one that needs to be kept by the herd owner. Lactation summaries, calving and breeding records and calf identification would be the responsibility of the official recorder.

A summary based on the data collected on the test day should be sent to the farmer shortly after the test to enable him to take the necessary management decisions. These should include (i) the test-day lactation report showing each animal's milk production and total production to the last test day and (ii) a herd summary giving information on animals for breeding, pregnancy checking, drying and culling during the next 30 days. The latter report should also give the means of milk yield, composition and calving intervals of animals that completed lactation during the last 12 months.

Beef Recording

Performance recording in beef cattle even in countries

with advanced beef industries has not attained the same degree of uniformity and refinement as that of the dairy industry. This is due to the diversity of environments and management systems under which beef cattle are raised and the different production systems used for beef production. The management systems under which beef animals are maintained in the tropics are even more variable. With the exception of the small proportion of cattle kept in commercial ranches, the majority of cattle are raised under extensive systems and are not accessible for recording. Records taken in the latter animals may have to be limited to traits that can be obtained by questioning the owners, e.g. reproductive traits, mortality.

Official recording schemes for collection of beef cattle records are in operation in many developed countries. Examples are the Meat and Livestock Commission (MLC) in the United Kingdom, Performance Registry International in USA, Record of Performance (ROP) in Canada and Beefplan in New Zealand. These organisations obtain records of pedigree, growth rate of calves and calving difficulties from private farmer herds.

In addition, carcass data are recorded on animals slaughtered. The home recording scheme as operated by the MLC involves weighing calves in private herds four times a year by MLC staff till the animals are 20 months old. The farmer pays a fee for this service, the amount paid depending on the size of the herd and whether the MLC weighbridge or the farmer's scale is used. It also operates a Central Performance Testing scheme where bulls from different herds are recorded together at a central station under standard management. Measurements recorded are weight changes, feed intakes to calculate feed conversion efficiency, withers height as a measure of stature and backfat thickness using a "scanogram" as a crude measure of carcass merit. Bulls are also scored for type at the end of the test.

Traits Recording

Since recording for beef requires more effort on the part of the producer than milk recording, it is necessary to ensure that he sees tangible benefits in return. The traits that should be recorded are as follows:

i. *Reproductive performance*: Fertility in beef cattle is usually measured in terms of calving rate, i.e. the number of calves born in relatioı to the number of mature females exposed to bulls. This estimate is generally adequate for survey studies and is easily obtainable where calving dates are collected over a period of time. It may not be adequate for management purpose since it does not reveal the extent of fertility disorders until actual calving occur. Where calving rate is to be deduced from an examination made on one visit, it can only estimated indirectly. Such estimates have been made in tropical beef cattle, e.g. in Colombia and Botswana, but they involve pregnancy testing and estimation of age of foetuses. For selection purpose, individual cow data are required for which the parameter usually used is the calving interval. Where records on calving dates are not obtainable, the only record that may be useful is the previous parous state of the cow, i.e. whether or not she had a calf during the year.

ii. *Mothering ability*: The ability of the dam to wean a heavy calf is important in beef cattle, particularly in cow-calf operations. It involves measurement of two traits—viability of the calf and weaning weight. Viability is simply a record of whether a calf died before weaning age and the cause of death. Ideally, weaning weight should be taken at a constant age between 180 and 240 days. However, for ease of management, calves are generally weaned in groups, each group consisting of animals born during a particular fortnight or month. In such situations, date of weaning should also be recorded to adjust weaning weight for age at weaning.

iii. *Post-weaning weight*: A sufficiently long time should elapse after weaning before post-weaning weight is taken to allow for adjustments caused by compensatory growth mechanisms and to permit animals to exhibit their own genetic potential. Weights are usually taken at some given end-point such as 12, 18 or 24 months of age depending on the management system. Where animals are recorded in test stations, the measurements obtained are weight changes from the beginning to end of test. There are also situations where calves are raised on range and finished in feedlot, e.g. feedlot operations in Kenya. In such circumstances, though weight changes in the two systems may be recorded for management decisions, the average daily gain at the end of the finishing period, which includes the components of weight gains on rangeland and during finishing, is adequate for selection purposes.

iv. *Other weights*: Birth weight is of economic importance in breeds prone to dystocia. In other breeds, it provides an accurate measure of liveweight gain from birth to weaning. Dam weight at parturition is of use in adjusting weaning weight while weight change of the dam from parturition to weaning is an indication of the ability of the cow to maintain weight during lactation. The recording of these weights is difficult under field conditions and may therefore be omitted except in institutional herds.

v. *Tick resistance*: The susceptibility of animals to tick infections is highly heritable. It is therefore an advantage to measure tick resistance in areas where ticks are endemic. The important points to be observed in measuring tick resistance are as follows: (a) Assessments of tick burden should be done when infestation is likely to be high, which is during the wet season; a population mean of at least 25 ticks per side of animal is considered desirable to reveal the genetic variation between animals. (b) Animals to be compared should have been together for at least 6 months. The assessment is usually made on the basis of two counts of the number of replete female ticks (4.5-8.0 mm in size for Boophilus microplus) after exposure to either natural or artificial infestation. This method is suitable for one-host ticks (Boophilus spp) and may not be applicable to other types.

vi. *Longevity*: Longevity in terms of length of productive life in a herd is important in the tropics since animals reach puberty later in life and therefore the longer they remain productive, the lower is the cost of raising a heifer per year of productive life. Animals with a long productive life should be identified so that their close relatives could be selected. Simple procedures such as ear notching could be used even in herds that are not recorded for production traits.

vii. *Coat type*: Coat type is a good indication of tropical adaptation in temperate breeds or in zebu x temperate crosses. Coat-type classification is made on the basis of a subjective score in which the primary criterion is the length of hair or depth of coat, with "handle" (a subjective assessment of diameter of hair) a modifying factor. Usually, the coat on the midside is scored by moving the palm of the hand over the side. Scores range from 1 to 7, 1 representing an extremely short type as found in zebus and 7 a very wooly coat.

viii. *Colour and conformation*: Colour is of little economic significance but gross deviations from the established norm warrant recording for possible culling since such animals may not be saleable as breeding stock. Visual assessment of conformation is a poor estimator of carcass quality in terms of muscle to bone ratio. It is merely indicative of condition which is a measure of fat cover. Therefore use of condition scores to reject animals selected on other economic criteria should be avoided.

ix. *Carcass quality*: Accurate estimation of carcass quality involves slaughter of animals. Practical and efficient methods for estimating carcass value in the live animal are still in the development phase. Hence this trait may be ignored in recording.

Field recording for beef production in the tropics requires more organisation and elaborate arrangements since weighbridges are usually not available in private herds. The method of recording given below in respect of weights and classification is being used successfully in Kenya and Botswana in large herds. It is not of value in sedentary herds with a small number of cattle or in nomadic herds which may not be available for recording when required.

i. *Identification*: The procedure adopted should give each animal a unique identification. Tamperproof metal ear tags are the best as they can be applied at birth and loss rates are low. For rapid identification in the field and also to serve as a back-up in case of loss, the tag number should also be branded on the upper thigh by hot iron at weaning.

ii. *Classification of animals*: When a herd is recorded for the first time, the animals are classified into one of two groups—breeding cows and other stock. The former includes females of breeding age and weight while the latter includes steers and bulls. Pedigree relationships should also be established at the first recording. Thus before an animal leaves the crush, its relationship with any other animal in the herd is determined. If it is a cow, information on whether she has a calf is obtained and similarly, if it is a calf, its dam number is obtained.

iii. *Records*: For breeding cows, records taken are: breed, date of birth (actual or determined approximately from dentition), sire and dam if known, previous parous state and weight. Similarly, for other stock, records of breed, date of birth, sex (male, female or castrate), sire and dam number, dam's age and its previous parous state and weight are taken.

iv. *Weight*: Weights are obtained monthly using either permanent or mobile weighbridges of (a) calves weaned during the month and (b) animals in the age range of 18-24 months. The ranch is informed in advance of the date of weighing so that animals to be weighed can be assembled the previous evening and kept off feed and water.

In addition to the above records which are taken by the central recording office, the ranch keeps separate books for birth records, weight records and treatment records. The latter record covers all information on breeding (services, etc.), health, management (weaning, castration), feeding (supplementary feeding) and stock changes (purchases, sales, etc.). The weight-record book is simply a record of the date of weighing and weights of the two group of animals—weaned stock and those 18-24 months of age.

Recording Sheep and Goats

Systems of sheep and goat breeding in developed countries have not shown as much change as those for dairy cattle, partly because important characteristics such as fleece or body weight can be seen or measured in both sexes and also because these animals are associated with less' favourable environments. One exception is the introduction of cooperative breeding schemes for sheep in New Zealand and Australia in recent years; these involve selection on measured performance.

Sheep and goats can produce meat, fibre and milk. However, some breeds (e.g. Merino sheep and Angora goats) are kept primarily for their fleece, others for milk (e.g. the sheep breeds producing Roquefort cheese in France, or European goat breeds such as the Saanen), others again mainly for meat.

Traits Recording

a. *Reproduction*: Reproductive rate is important in all systems of sheep production. An increased rate lowers overhead maintenance costs for ewes, allows an increased selection differential for productive traits and gives more surplus animals for sale. The measurement of economic significance is the number of lambs weaned per ewe joined per year which takes into account both lambing rate and lamb survival rate. In addition to litter size at birth, the New Zealand "Sheeplan" (national recording system) records weight of lamb weaned per ewe, which places emphasis on number of lambs weaned rather than number born, and so includes survival rate.

b. *Mortality*: Mortality at all ages, but particularly from birth to weaning, is a serious problem in the sheep industry and should be recorded. Survival rates of lambs to weaning are usually lower for multiple-born than single-born lambs, but in developed countries are seldom less than half, so that multiple births still give more lambs at weaning. Adult mortality can also be as high as lamb mortality in the tropics and should therefore be recorded.

c. *Birth weight*: The only reason for recording birth weight would be to make it possible to calculate growth rate to weaning, but this is seldom worthwhile. It is usually sufficient to record weaning weight at a known age.

d. *Milk yield*: Milk yield accounts for most of the variation in early growth of lambs. Where lamb growth is to be used as a measure of maternal milk yield, growth over the first 4-5 weeks is a good indicator as the correlation between milk yield and lamb growth at this stage is high. In dairy breeds of sheep and goats, direct estimation of milk yield as in dairy cattle is often not possible as the dam generally suckles the young and milking and suckling are concurrent. She-goats in particular are temperamental and may refuse to let down milk if separated from their kids.

e. *Growth rate*: The end product of meat breeds of sheep raised in advanced sheep-raising countries is the weaned lamb. Weaning weight is thus an important trait for performance recording, though Young et al. (1965) showed that weight at 15-16 months of age in Merinos had a higher heritability and a high genetic correlation with weaning weight. In the tropics, animals are rarely slaughtered at weaning, so body weights at later ages with their high heritability are more important than weaning weight.

f. *Wool*: Wool is an important aspect of sheep production in some areas of the tropics and with this product end use must be carefully considered. The main division is into apparel and carpet wools, the most important traits for both being:

i. *Quantity measured by wool weight per head*: The required weight is the clean weight, but greasy weight is simpler to measure and has a high heritability and a high genetic correlation with clean weight.

ii. *Quality, the important traits being average fibre diameter, staple length, percent medullated fibres, kemp and fibre strength*: Besides these main points, consideration should also be given to market requirements. Thus, hand processing, e.g. cottage industries, cannot deal with fine wools, particularly in the scouring and spinning stages. This is important for apparel wools, but not for carpet wools. In wool meant for world markets or for machine processing in the country of origin, the standards are usually laid down by the users.

Recording Procedures

Where records are used as an adjunct to the management of commercial flocks, individual identification is not necessary, as flock records are adequate. The minimum records necessary for this purpose are costs of supplementary feed and labour and sale value of animals, milk and wool. When information on the performance of the flock is required, periodic surveys may be made to gather data on the number of lambs weaned per ewe and mortality rates of lambs and older animals.

The procedures adopted in recording schemes for genetic improvement very with the objectives and capabilities of each flock. A flexible system of recording can be used, starting with minimal records that could be made elaborate with progress in record keeping. The type of records are set out below:

a. *Identification*: Some effective but less accurate selection procedures are possible even without individual identification. Where it is necessary to carry out more accurate selection procedures, permanent identification is achieved by the use of metal or plastic tags, tattooing on the ear or tail and horn branding.

b. *Pedigree records*: Dam identification is adequate for simple selection procedures. It should be done within a few hours after lambing by pairing ewes and lambs. Where identity of the sire is required, it is necessary to resort to hand mating or to run each ram with a group of females in fenced paddocks.

c. *Lambing and weaning record of ewe*: The lambing record of the ewe should indicate whether she was barren or produced single or multiple-born lambs and the number of lambs she raised to weaning age. The date of lambing, and the sex and identification of each lamb should be stated. In the flocks that lamb seasonally, lambing records should be preferably be taken daily. However, in many tropical countries, the rams run with the females throughout the year, so that there is no definite lambing season. Recording of births may then be done fortnightly or monthly. In flocks that are not individually identified, ewes which fail to lamb during each 12-month period may be given ear notches to aid culling of barren ewes.

d. *Weaning weight*: Since lambs are usually weaned in groups, all animals in a group are weighed on the day of weaning. Even where weaning is staggered, weighings could be`done on predetermined dates and data adjusted for age differences.

e. *Post-weaning weight*: This is usually taken at any age between 12 and 18 months of age. Where type of birth and age of dam are not recorded, the weight should be taken at an age close to 18 months when the effects due to these factors are minimal.

f. *Wool*:

i. *Quantity*: For sheep shorn once a year, records of wool weights at 13-18 months of age, following a previous shearing at weaning, are a good indication of lifetime production. Sheep in many tropical countries, however, may be shorn twice or even three times a year. The sum of the weights at the second and third shearings should be recorded for sheep that are shorn twice a year, and at the second, third and fourth if shorn three times. Accurate records of the time between shearings should also be kept. Greasy weight is adequate in most cases but, for Merino types, it may be desirable to measure percent clean yield.

ii. *Quality*: Fineness in apparel wools has traditionally been estimated through the number of crimps per inch of staple. However, this is an inaccurate guide and it has therefore been replaced in Australia by measurements taken on a sample from the midside. The measurements include fibre diameter, percent clean yield and amount and type of vegetable matter.

g. *Milk yield*: Indirect methods are used in estimating milk yield to overcome the difficulties caused by suckling. One method involves separation of the young from its dam for a 24-hour period on recording days which are usually one month apart. Another method is based on the liveweight increase of the lamb from birth to 2 months of age. This is multiplied by a conversion factor to estimate the amount of milk sucked which is then added to the amount milked out.

h. *Disease*: Records of ill-health and their causes are also sometimes recorded.

Bibliography

Adams H., "Improved Seed Production: Caricom Countries and Suriname", East Coast Demerara, Guyana: CARDI, 1993.

Alarcon E., "Plant genetic resources in Latin America and the Caribbean: an institutional overview". San Jose, Costa Rica: *Inter-American Institute for Cooperation and Agriculture,* 1998.

Ayad W G, Hadgkin T, Jaradat A and Rao V R (eds.), "Molecular genetic techniques for plant genetic resources", *In: Proceedings of Meeting held at Rome,* Italy, 9-11 October 1995.

Barton J. H. and Siebeck W E., "Material transfer agreements in genetic resources exchange: the case of the International Research Centres", Rome, Italy: International Plant Genetic Resources Institute, 1994

Bateson J.M, McDonald F.D and Robertson N.E., "Need for economic assessment in biotechnological development", St Augustine, Trinidad and Tobago: CARDI. 1992.

Benson E.E., "Free radical damage in stored plant germplasm", Rome, Italy: IPGRI. 1990.

Broomes V.F.A., "Cultivation of tissue culture plants by farmers in Guyana: a review of activities", *Proceedings of annual review conference 1992,* Mon Repos, Guyana, 20-23 October 1992.

Cropper D, Engles J and Frison E., "A multilateral system for plant genetic resources: imperatives, achievements and challenges", *Issues in Genetic Resources*; No. 2 Rome, Italy: International Plant Genetic Resources Institute. 1994.

Duvick D N., "Plant breeding and biotechnology for meeting future food needs". *In: Population and food in the early twenty-first century: meeting future food demands of an increasing population.* Washington DC, USA: International Food Policy Research Institute,1995.

Food and Agriculture Organisation of the United Nations, "Report on the state of the world's plant genetic resources for food and agriculture". *International Technical Conference on Plant Genetic Resources,* Leipzig, Germany 17-23 June 1996.

Guarino L, Ramanatha Rao V. and Reid R. "Collecting plant genetic diversity: technical guidelines", Oxon, UK: CAB International, 1995.

Hardon J., "Participatory plant breeding", *Meeting held at Wageningen,* The Netherlands, 26-29 July 1995.

Hodgkin T, Brown A.H.D, Hirtum J.L van and Moracis E.A.V (eds.)., "Core collections of plant genetic resources", New York, USA: John Wiley, 1995.

Murray D.R (ed.), "Advanced methods in plant breeding and biotechnology", Wallingford, UK: CAB International, 1991

Index